Quantum Reality

Quantum Reality

BEYOND THE NEW PHYSICS

Nick Herbert

ANCHOR PRESS/DOUBLEDAY
Garden City, New York
1985

Library of Congress Cataloging in Publication Data
Herbert, Nick.
Quantum reality.
Includes index.
1. Quantum theory—Popular works. I. Title.
QC174.12.H47   1985   530.1'2   82-46033
ISBN 0-385-18704-1

To my father

Contents

Preface

One of the curious features of modern physics is that in spite of its over-whelming practical success in explaining a vast range of physical phenomena from quark to quasar, it fails to give us a single metaphor for how the universe actually works. The old mechanical metaphor "The world is a giant clock" condensed in one image the principal features of Newtonian physics—namely, atomicity, objectivity, and determinism. However, physicists today do not possess a single metaphor that unites in one image the principal features of quantum theory. The main purpose of *Quantum Reality* is to examine several tentative images of the world proposed by quantum physicists.

The search for a picture of "the way the world really is" is an enterprise that transcends the narrow interests of theoretical physicists. For better or for worse, humans have tended to pattern their domestic, social, and political arrangements according to the dominant vision of physical reality. Inevitably the cosmic view trickles down to the most mundane details of everyday life.

In the Middle Ages, when virtually everyone believed the world to be the personal creation of a divine being, society mirrored the hierarchy that supposedly existed in the heavens. Dante's picture of this world as a series of concentric spheres—heaven the largest; next, the planets' crystalline spheres; down through our Earth's concentric "elements," the whole supported by the seven circles of hell—gave everything and everyone his proper place in the medieval scheme of things, from the divine right of kings down to the abject obedience of the lowliest serf. Most people accepted this hierarchical structure without question because it represented the way the world really is.

The Newtonian revolution toppled the reign of the crystal spheres and replaced it with a physics of ordinary matter governed by mathematical laws rather than divine command. Coincident with the rise of Newtonian physics was the ascent of the modern democracy which stresses a "rule of laws rather than men" and which posits a theoretical equality between the parts of the social machinery. The Declaration of Independence, for example, "We hold these truths to be self-evident" reads more like a mathematical theorem than a political document. As above, so below. The egalitarian mechanism that Newton discovered in the heavens has insinuated itself into every aspect of ordinary life. For better or for worse, we live today in a largely mechanistic world.

Just as Newton shattered the medieval crystal spheres, modern quantum theory has irreparably smashed Newton's clockwork. We are now certain that the world *is not* a deterministic mechanism. But what the world *is* we cannot truly say. The search for quantum reality is a search for a single image that does justice to our new knowledge of how the world actually works.

Many aspects of quantum theory are public knowledge, such as the notion that all elementary events occur at random, governed only by statistical laws; that there is a "least thing that can happen"—Max Planck's irreducible constant of action; and that Heisenberg's famous uncertainty principle forbids an accurate knowledge of a quantum particle's position and momentum. A successful quantum reality would incorporate this knowledge, and much more, into a single comprehensive metaphor for the way the world really is.

I first encountered the quantum reality question in graduate school when I learned to describe the behavior of atoms, molecules, and elementary particles in the mathematical language of quantum theory. Quantum theory is peculiar in that it describes *a measured atom* in a very different manner than *an unmeasured atom.*

The measured atom always has definite values for its attributes (such as position and momentum), but the unmeasured atom never does. Every atom in the world that's not actually being measured possesses (in the mathematical description at least) not one but all possible attribute values, somewhat like a broken TV set that displays all its channels at the same time.

Of course I wondered what sort of reality this strange symbolization of the unmeasured world actually stood for. Were the attributes of unmeasured atoms multivalued, fuzzy, nonexistent, or simply unknown?

However, when I asked my teachers what quantum theory actually meant—that is, what was the *reality* behind the mathematics—they told me that it was pointless for a physicist to ask questions about reality. Best to stick with the math and the experimental facts, they cautioned, and stop worrying about what was going on behind the scenes. No one has expressed physicists' reluctance to deal with quantum reality better than Richard Feynman, a Nobel laureate now at Cal Tech, who said, "I think it is safe to say that no one understands quantum mechanics. Do not keep saying to yourself, if you can possibly avoid it, 'but how can it be like that?' because you will go 'down the drain' into a blind alley from which nobody has yet escaped. Nobody knows how it can be like that."

For the sake of having *something* in mind while I did my quantum calculations, I imagined that an atom always possessed definite values for all its attributes (just like an ordinary object) whether that atom was measured or not. However, the process of measurement disturbs the atom so profoundly that its measured attributes bear only a statistical relation to its unmeasured attributes. I felt sure that such a "disturbance model" of measurement was capable of accounting for quantum randomness, the Heisenberg uncertainty relations, and other quantum mysteries as well. In this "disturbance" picture, an atom's *actual* position and momentum are always definite but usually unknown; its *measured* position and momentum cannot be accurately predicted because the measuring device necessarily changes what it measures.

My belief in this disturbance model of reality was strengthened when I read that young Werner Heisenberg once held a similar view of the quantum world. It did not occur to me to wonder why Heisenberg quickly abandoned such an obvious explanation to take up the more obscure and mystical Copenhagen interpretation, which most physicists endorse today.

In brief, the Copenhagen interpretation holds that in a certain sense the unmeasured atom is not real: its attributes are created or realized in the act of measurement.

I regarded the Copenhagen interpretation as sheer mystification compared to the clarity and common sense of my disturbance model. Blissfully ignorant concerning the real issues surrounding the quantum reality question, I got my degree and continued my career as an industrial and academic physicist.

In the summer of 1970 my friend Heinz Pagels, a physicist at Rockefeller University, showed me a paper published in an obscure new journal. "Here's something strange that should interest you, Nick," he said. This strange new thing was Bell's theorem, a mathematical proof which puts strict conditions on any conceivable model of reality, quantum or otherwise.

Bell's theorem is easy to understand but hard to believe. This theorem says that *reality must be non-local.* "Non-local" means, in terms of my disturbance model, that the atom's measured attributes are determined not just by events happening at the actual measurement site but by events arbitrarily distant, including events outside the light cone—that is, events so far away that to reach the measurement site their influence must travel faster than light. In other words, when I probe an atom's momentum with a momentum meter, its true momentum is disturbed, according to Bell's theorem, not just by the momentum meter itself but by a vast array of distant events—events that are happening right now in other cities, in other countries, and possibly in other galaxies. According to John Bell, the act of measurement is not a private act, but a public event in whose details large portions of the universe instantly participate.

Bell's theorem is a mathematical proof, not a conjecture or supposition. That is, once you accept a few simple premises his conclusion certainly follows. Thus Bell does not merely permit or suggest that reality is non-local; he actually proves it.

Bell's theorem has immensely clarified the quantum reality question. For instance we now know for certain that no local model (such as my naïve disturbance model) can explain the quantum facts. Bell's theorem has important consequences for all models of quantum reality including the Copenhagen interpretation, and its effects continue to reverberate in physics circles. This book explores various quantum realities (models of the world consistent with quantum theory) in the light of Bell's important discovery.

Many people have helped me in my search for quantum reality, either through their books and articles or through personal contact. I can mention only a few but I'm grateful to all.

I would like to honor the memory of Randy Hamm, friend and talented animator whose collaboration on *Benjamin Bunny Faces Reality,* an unfinished animated film which explores some of the same concepts contained in this book, inspired me to think in new directions.

I would like to thank Mike and Dulce Murphy for opening Esalen Institute, Big Sur, to physics conferences on quantum reality. Thanks also to the many participants in the Esalen conferences, especially Henry Stapp, Saul-Paul Sirag, John Clauser, David Finkelstein, John Cramer, Larry Bartell, H. Dieter Zeh and Bernard d'Espagnat, from whom I received much enlightenment concerning the quantum mysteries.

Thanks to Charles Brandon and The Reality Foundation for encouragement and a timely graphics grant, to Lynn Miller for her skillful illustra-

A frame from Benjamin Bunny Faces Reality: *the "Professor" readies Benjamin for a "reality check."*

tions, to Shirlee and David Byrd for editorial assistance, and to Doubleday's Phil Pochoda, Dave Barbor, and Chaucy Bennetts for their patience and good advice.

Thanks to my wife Betsy and son Khola for keeping me awake and aware of other extraordinary realities.

Quantum Reality

1

The Quest for Reality

The essential point in science is not a complicated mathematical formalism or a ritualized experimentation. Rather the heart of science is a kind of shrewd honesty that springs from really wanting to know what the hell is going on!

Saul-Paul Sirag

When I was six my parents gave me a set of children's books—fourteen orange, black, and gold bound volumes of stories, games, and songs. *Science* was Volume 12, the only book without text, containing instead dozens of black-and-white photographs of big machines and unusual natural phenomena. One picture in particular fascinated me; recalling it today still makes me shiver. This picture showed a nest of eggs. But hatching out of these eggs were baby snakes.

This disturbing photo brought together in one image my vague fears that beneath the surface of commonplace things lurks an utterly strange (and probably sinister) reality.

Many years later I experienced that same feeling—a lightning realization that this world is not what it seems—precipitated not by a picture in a children's book but by a mathematical argument in a physics journal. Bell's theorem is a simple but powerful proof concerning the structure of physical reality, and had the same effect on my imagination as that snake's nest. Bell's theorem is one of the clearest windows that physicists possess into the nature of deep reality. I invite you in Chapter 12 to look through this window too.

Physicists are interested in how the world is put together—out of what sorts of basic objects, interacting via what sorts of basic forces. Physics began in antiquity as a kind of natural history, a folk museum of unexplained marvels and peculiar facts laid out in haphazard fashion: the world as *lore*, direct observation scrambled up with fantastic travelogue, with medieval bestiaries and alchemical recipes.

In the seventeenth century Galileo, Newton and other natural philosophers discovered that an enormous body of physical facts could be encompassed in a few mathematical formulas. For instance with only three mathematical laws Newton could explain all motion in heaven and on Earth. Why should mathematics, developed primarily to keep track of human business transactions, have anything at all to do with the way the non-human world operates? Nobel laureate Eugene Wigner refers to this magical match between human mathematics and non-human facts as "the unreasonable effectiveness of mathematics in the natural sciences." "This unreasonable effectiveness," writes Wigner, "is a wonderful gift which we neither understand nor deserve."

Although mathematics originates in the human mind, its remarkable effectiveness in explaining the world does not extend to the mind itself. Psychology has proved unusually resistant to the mathematization that works so well in physics.

The German philosopher Immanuel Kant was deeply impressed by Newton's mathematical method and sought to explain its success as well as to understand its limitations. Kant began his analysis by dividing knowledge into three parts: appearance, reality, and theory. Appearance is the content of our direct sensory experience of natural phenomena. Reality (Kant called it the "thing-in-itself") is what lies behind all phenomena. Theory consists of human concepts that attempt to mirror both appearance and reality.

Kant believed that the world's appearances were deeply conditioned by human sensory and intellectual apparatus. Other beings no doubt experience the same world in radically different ways. Scientific facts—the appearances themselves—are as much a product of the observer's human nature as they are of an underlying reality. We see the world through particularly human goggles. Kant felt that the participation of human nature in the creation of appearances explained both the remarkable ability of human concepts to fit the facts and the natural limits of such abilities.

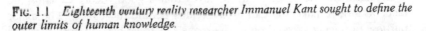

Fɪɢ. 1.1   *Eighteenth century reality researcher Immanuel Kant sought to define the outer limits of human knowledge.*

Our concepts appear to match the facts, according to Kant, because both facts and concepts have a common origin—the human condition. Insofar as human nature is entwined with the appearances, human concepts will be successful in explaining those appearances. Because we can only explain those aspects of the world which we ourselves bring to it, the nature of deep reality must remain forever inaccessible. Man is fated to know, either directly or through conceptualization, merely the world's appearances and of these appearances only that part which is of human origin.

Kant's position is an example of the pessimistic pole of reality research, which might be expressed this way: human senses and intellectual equipment evolved in a biological context concerned mainly with survival and reproduction of humankind. The powers that such clever animals may possess are wholly inadequate to picture reality itself, which belongs to an order that utterly transcends our domestic concerns.

On the other hand, reality researchers of an optimistic bent argue that since humans are part of nature, deeply natural to the core, nothing prevents us from experiencing or conceptualizing reality itself. Indeed some of our experiences and/or some of our ideas may already be making contact with rock-bottom reality.

Besides the optimism/pessimism split, another difference separates researchers into the nature of reality: the pragmatist/realist division. A pragmatist believes only in facts and mathematics and refuses in principle to speculate concerning deep reality, such questions being meaningless from his point of view. Sir James Jeans, the distinguished physicist and astronomer, sums up this pragmatic orientation: "The final truth about a phenomenon resides in the mathematical description of it; so long as there is no imperfection in this, our knowledge of the phenomenon is complete. We go beyond the mathematical formula at our own risk; we may find a model or picture which helps us understand it, but we have no right to expect this, and our failure to find such a model or picture need not indicate that either our reasoning or our knowledge is at fault. The making of models or pictures to explain mathematical formulas and the phenomena they describe is not a step towards, but a step away from, reality; it is like making graven images of a spirit."

A realist, on the other hand, believes that a good theory explains the facts because it makes contact with a reality behind those facts. The major purpose of science, according to the realists, is to go beyond both fact and theory to the reality underneath. As Einstein, the most famous realist of them all, put it, "Reality is the real business of physics."

The pragmatist treats his theory like a *cookbook* full of recipes which are useful for ordering and manipulating the facts. The realist sees theory as a *guidebook* which lays out for the traveler the highlights of the invisible landscape that lies just beneath the facts.

Most physicists are complex mixtures of pragmatist and realist, at once both optimistic and pessimistic about their chances for making solid contact with deep reality. Like many other human enterprises, the practice of science requires a constantly shifting balance between extremes, a sensitivity to the middle way, as French experimentalist Jean Perrin put it, "between the instincts of prudence and audacity necessary to the slow progress of human science."

I do not wish to get specific about what might be meant by "reality itself" lest we hamper our search with needless preconceptions. Like the solution to a puzzle or cryptogram, contact with deep reality will no doubt carry its own validation: we'll know it when we see it. However, as an illustration of the kinds of realities we might be looking for in physics, I review here two case histories: the stories of a reality that failed and a reality that succeeded.

THE LUMINIFEROUS ETHER

In 1864 Scottish physicist James Clerk Maxwell discovered the basic equations that govern electricity and magnetism. To everyone's surprise these phenomena turned out to be two aspects of a single entity—the electromagnetic field. Today physicists are seeking for a way to unify all of nature's fields. Maxwell was the first physicist to show that the task of field unification is not futile.

A theoretical bonus which Maxwell reaped from his fusion of two fields into one was the discovery that waves in his electromagnetic field traveled at the same speed as the measured velocity of light. On the basis of this numerical coincidence, Maxwell conjectured that light, in reality, was an electromagnetic vibration at a particular frequency. The experimental production by Heinrich Hertz of low-frequency electromagnetic radiation (radio waves) which were identical in all respects save frequency to light confirmed Maxwell's bold conjecture.

All known waves vibrate in some medium (such as air or water). The medium in which light presumably travels was dubbed the "luminiferous ether." Late nineteenth-century physicists gave top priority to research into the ether's mechanical properties. Maxwell described the subject of this research in these words: "Whatever difficulties we may have in forming a consistent idea of the constitution of the ether, there can be no doubt that the interplanetary and interstellar spaces are not empty but are occupied by a material substance or body, which is certainly the largest, and probably the most uniform body of which we have any knowledge."

From light's well-known attributes one could infer many of this hypothetical ether's properties. For instance, since light travels so fast the elasticity of the ether must be enormous, millions of times more resilient than the hardest spring steel. Since light is a transverse wave—vibrating sidewise rather than back and forth—the ether must be a solid. Gases and liquids can support only back-and-forth vibrations (sound is an example), while solids transmit both kinds of vibration (earthquake waves, for example, vibrate in all directions). The fact that light vibrates only sidewise (no back-and-forth light has ever been observed) had to be explained by complex structures in the ether which suppressed altogether this otherwise natural back-and-forth vibration but which permitted sidewise vibrations to propagate with extreme rapidity.

Although the universe was filled with a transparent "glass" much harder than steel, this glass offered not the slightest resistance to the passage of material bodies. The Earth's motion was seemingly unaffected by the presence of the luminiferous ether. Some physicists proposed that the ether might act like a solid for rapid motions such as light, but like a fluid for slow motions, such as planets, in the manner of certain waxlike solids with deformation-rate dependent viscosities. In modern terms, such a hypothesis amounts to assuming that the universe is filled brim to brim with a kind of Silly Putty.

In 1887 two American physicists performed a simple experiment whose purpose was to determine the velocity of the Earth through this ever-present vibrating solid. Albert Abraham Michelson and Edward Williams Morley set up a kind of optical racetrack that pitted a light beam moving north and south between parallel mirrors against an east/west beam. Depending on the direction of the "ether wind," one or the other of these beams had the track advantage and was sure to win. The result of the Michelson-Morley experiment was always a photo finish. Despite the enormous velocity of the Earth through space, a velocity that constantly changes its direction during the year, the two experimenters could detect no movement whatsoever of the luminiferous ether past the Earth.

Michelson and Morley's failure to detect the "ether wind" led physicists to propose that massive bodies such as the Earth trap the ether and carry it along with them. However, attempts to detect this "ether drag" near massive rotating bodies in the laboratory were unsuccessful. Ether drag should also distort the apparent positions of distant stars, an effect which was also noticeably absent.

To explain the failure of Michelson and Morley to detect an "ether wind," even more preposterous effects were invoked. Dutch physicist Hendrik Antoon Lorentz and Irish physicist George Francis FitzGerald independently proposed that motion through the ether resulted in a tiny contraction of all physical bodies in the direction of motion. The Lorentz-FitzGerald contraction—a kind of "ether squeeze"—could not be directly observed, because measuring rods also supposedly shrank when oriented in the ether wind's direction. The sole function of the Lorentz-FitzGerald contraction was to even the odds in the Michelson-Morley optical race-track: the light beam that would have lost the race, by virtue of the L-F contraction would now travel a shorter path, and consequently both beams would reach the finish line at precisely the same time. This hypothetical "ether squeeze" was a desperate attempt to save appearances by loading the already peculiar ether with yet one more unusual attribute.

Although its properties grew more preposterous with each new investigation, the existence of the ether itself was never called into question. One of England's leading physicists, the eminent William Thomson, Lord Kelvin, expressed physicists' general attitude a few years after the Michelson-Morley experiment when he said, "One thing we are sure of, and that is the reality and substantiality of the luminiferous ether."

Despite physicists' strong faith in the existence of the luminiferous ether, a few years after Lord Kelvin's profession of belief the ether was swept away into the junkyard of obsolete physical concepts in company with phlogiston, caloric, and the fabled philosopher's stone.

Albert Einstein, an unknown clerk in the Zurich patent office, published in 1905 a new theory of space and time which came to be called the special theory of relativity. The cornerstone of Einstein's theory was that only *relative* motions were of any consequence for the basic laws of physics. According to Einstein, there is no physical means by which one can observe a body's *absolute* motion through space.

Einstein's innocent assumption had far-reaching experimental consequences. For instance, two observers in relative motion measuring the positions and times of the same events would get different results. For Einstein, time and space are *relative* concepts, different for every observer. Another important consequence of relativity is the famous $E = mc^2$ relation, which predicts that an object's mass is equivalent to a certain amount of energy, with an enormous conversion factor—the square of the speed of light.

Although space, time, and motion are relative concepts in Einstein's theory, certain other physical quantities are absolute—the same for everyone. Einstein's major insight, the key to relativity theory, is that all valid physical laws must be built from these absolute quantities alone. Only in this way can these laws be made the same for all observers.

One of Einstein's absolutes is the speed of light: it is the same for an observer on Mars as an observer on Earth. Another Einsteinian absolute is the so-called *space-time interval*. Although space and time by themselves are different for each observer, a certain mathematical combination of space and time, chosen so that changes in space cancel the changes in time, is the same for everyone. Relativity's intimate linkage of space and time in the invariant space-time interval gives rise to the notion that in reality the world is *four-dimensional*—consisting of three spatial dimensions and one time dimension. The space-time interval is a kind of "distance" in that four-dimensional space.

According to Einstein, only such absolute quantities can be used as the ingredients of a valid physical law. Laws built to Einstein's specifications are called "covariant." Today we know for sure that if a physical theory does not have a covariant formulation it cannot represent the facts. By looking at where it's not covariant—wherever it uses a relative concept rather than an absolute one—we can even predict exactly how it must fail.

The luminiferous ether—a body that's "standing still in space"—is a manifestly non-covariant concept because it is standing still for only one observer. According to Einstein, the physics governing the interaction of bodies A and B can depend only on their relative motion, not on their velocity measured with respect to some special reference frame. If Einstein's theory is correct—and it has been abundantly verified—the concept of the ether can never enter into any correct physical law. The ether is quite literally a useless concept: there is no use for it in physics. However a light wave travels through space—light's manner of propagation is still a bit of a mystery—it cannot go via a medium made of invisible Silly Putty which fills up the universe.

Despite its central role in Victorian science, the luminiferous ether plays no part whatsoever in modern physics. The ether is a reality that failed. We consider next a reality that succeeded—the notion that matter is made out of atoms.

THE ATOMICITY OF MATTER

The idea that the world consists of standard little parts originated in antiquity. It would be hard to find a more eloquent statement of the atomic hypothesis than that of Democritus of Abdera, who wrote (about

500 B.C.): "By convention sour, by convention sweet, by convention colored; in reality, nothing but Atoms and the Void."

The atomic hypothesis existed side by side with the belief that the world consisted of transformations of a single continuous substance which some called "Fire," others "Air" or "Water". The commonplace observation that water could take solid, liquid, or gaseous forms depending on temperature was taken as an example of how one seamless substance might be able to simulate the world's enormous variety. However, until the nineteenth century the arguments for the continuum and the atomic hypotheses were mainly rhetorical; little evidence existed for either of these views.

In 1808 the British chemist John Dalton discovered that chemical substances combine according to fixed ratios—one part oxygen, for example, combines with two parts hydrogen to make water, provided each of these parts is assigned a standard weight. The standard weight of oxygen is sixteen times the standard weight of hydrogen. Dalton proposed that these constant combining ratios represented the combination of actual atoms whose atomic weights were proportional to the standard weights. According to Dalton, bulk hydrogen combines with bulk oxygen in a two-to-one ratio because water, in reality, is composed of two hydrogen atoms plus an oxygen atom. Dalton took these constant chemical ratios as tokens of an invisible atomic reality.

Most scientists were convinced by Dalton's arguments and accepted the real existence of atoms as an explanation of chemical reactions. However, a small but prestigious minority opposed the atomic hypothesis on the grounds that it went beyond the facts.

In 1826 Dalton received the Royal Society of London's medal of honor from famous British chemist Humphry Davy. While celebrating the importance of Dalton's work, Davy cautioned that the word "atom" could only realistically have the meaning "chemical equivalent"—that the atom was a unit of chemical reaction rather than a material entity. Davy praised Dalton for his discovery of the law of chemical proportions and predicted that his fame would rest on this practical discovery rather than on his speculations concerning invisible entities behind the phenomena.

Chemists of diverse nationalities united to oppose the atomic hypothesis. For instance, the distinguished French chemist Jean Baptiste Dumas proclaimed: "If I were master of the situation, I would efface the word atom from Science, persuaded that it goes further than experience, and that in chemistry, we should never go further than experience." The German chemist Kekulé, famous for his discovery of the benzene ring (which he presumably interpreted purely symbolically) had this to say about atoms: "The question whether atoms exist or not has little significance from a chemical point of view; its discussion belongs rather to metaphysics. In chemistry we have only to decide whether the assumption of atoms is a hypothesis adapted to the explanation of chemical phenomena."

"And who has ever seen a gas molecule or an atom?" chided Marcelin Berthelot, expressing the disdain that many of his fellow chemists felt for invisible entities inaccessible to experiment. Even its defenders saw little hope of ever directly verifying the atomic hypothesis: the size of these elementary entities—if they were really there at all—was estimated to be thousands of times smaller than a wavelength of light, hence technically forever invisible.

Wilhelm Ostwald, a German chemist who later received the Nobel Prize, turned to the field of chemical thermodynamics for an alternative to the atomic hypothesis. The two laws of thermodynamics—which require conservation of energy and an entropy-based limit on this energy's utilization—had been extended by Maxwell and Gibbs to describe successfully the intimate details of physical and chemical reactions without recourse to the atomic hypothesis. The success of the thermodynamic approach convinced Ostwald and his followers that molecules and atoms were imaginative fictions and that the real underlying component of the universe was energy, in its various forms.

Because of their faith in energy rather than atoms as an explanatory principle, Ostwald and his colleagues were called "energeticists." Debates in the journals and at scientific conferences between the supporters of the atomic hypothesis and the energeticists were sharp and emotional. The bitter opposition of the anti-atomists to his work on the kinetic theory of gases may have been partly responsible for the suicide of Ludwig Boltzmann, a brilliant but troubled theoretical physicist, in 1906.

In 1905, the same year he conceived the theory of relativity which demolished the luminiferous ether, Einstein published a paper on Brownian motion that pointed the way to conclusive experiments bearing on the real existence of atoms.

Whenever micron-sized particles are suspended in a liquid they undergo a perpetual quivering dance whose origin had been a mystery since its discovery in 1828 by Scottish botanist Robert Brown. Early experiments on Brownian motion were performed with pollen grains and the activity was believed to be of biological origin. I remember my first glimpse of what I took to be "cells" under a powerful microscope, and was fascinated by their ceaseless pulsations like tiny heartbeats until my teacher told me that I was looking at the Brownian motion of dirt particles. (Actually, when I finally spotted the real cells they didn't seem so interesting as the dancing dirt.) When it was discovered that any sort of finely divided matter would show such agitation (even stone from the Sphinx was pulverized and made to dance under the microscope), the biological hypothesis was discarded and various physical mechanisms proposed: temperature gradients, surface tension, obscure electrochemical effects. None of these quite worked. Brownian motion remained a minor mystery tucked away in an obscure corner of physics.

Einstein explained Brownian motion as the action of numerous atoms in motion colliding with the Brownian particle. This explanation had been previously rejected because the atoms were millions of times less massive than the Brownian particle, and their collective pressure could lead to no net motion because equal amounts of atoms were pushing in every direction.

Einstein showed that although the number of atoms striking the Brownian particle from each direction is equal on the average, the fluctuations away from this average lead to unbalanced forces in random directions. In any random process, the relative fluctuations from an average value is inversely proportional to the square root of the number of samples —the smaller the sample, the bigger the fluctuations. For a large particle, the bulk pressure of the surrounding atoms is indeed evenly balanced, but for a small particle, the fluctuations in the number of impinging atoms is sufficient to propel it in an unpredictable direction with a predictable force. Einstein showed how this random force would vary with temperature and particle size. If atoms existed, Einstein's model of Brownian motion would allow you actually to count the number of atoms striking a Brownian particle by measuring how far it drifts in response to these fluctuation forces.

In a series of ingenious experiments the French physicist Jean Baptiste Perrin verified Einstein's model and succeeded for the first time in actually counting the number of atoms in a drop of water. Perrin published his direct verification of the atomic hypothesis in 1913, in a book called simply *Les Atomes*.

In 1895 Ostwald railed against the atomic hypothesis in a speech entitled "On Overcoming Scientific Materialism": "We must renounce the hope of representing the physical world by referring natural phenomena to a mechanics of atoms. 'But'—I hear you say—'but what will we have left to give us a picture of reality if we abandon atoms?' To this I reply: 'Thou shalt not take unto thee any graven image, or any likeness of anything.' Our task is not to see the world through a dark and distorted mirror, but directly, so far as the nature of our minds permits. The task of science is to discern relations among *realities*, i.e., demonstrable and measurable quantities . . . It is not a search for forces we cannot measure, acting between atoms we cannot observe."

But in response to the work of Einstein and Perrin, the leader of the energeticists bowed to the experimental evidence and finally accepted the real existence of atoms: "I am now convinced," said Ostwald, "that we have recently become possessed of experimental evidence of the discrete or grained nature of matter, which the atomic hypothesis sought in vain for hundreds and thousands of years. [Experiments such as Perrin's] justify the most cautious scientist in now speaking of the experimental proof of the atomic nature of matter. The atomic hypothesis is thus raised to the position of a scientifically well-founded theory."

More recently (1957) philosopher of science Hans Reichenbach summed up the modern opinion concerning the atomic hypothesis: "The atomic character of matter belongs to the most certain facts of our present knowledge . . . we can speak of the existence of atoms with the same certainty as the existence of stars." The actuality of atoms is a reality that succeeded. Nobody today doubts that atoms really exist.

According to the pragmatists, science is like a cookbook—mere recipes for ordering phenomena. Once you have a recipe that works, what more could you ask for? Realists want more. They believe that a good theory should act as a guidebook to what's really out there in the world. In the words of Michael Polanyi, a distinguished scientific realist: "A theory which we acclaim as rational in itself is accredited with prophetic powers. We accept it in the hope of making contact with reality; so that being really true, our theory may yet show forth its truth through future centuries in ways undreamed of by its authors."

Quantum theory has been universally successful in describing phenomena at all levels accessible to experiment. It's a perfect cookbook, for whatever we choose to cook up. However, this comprehensive practical success has been accompanied by an unprecedented disagreement as to what quantum theory actually means, and a corresponding confusion as to what sort of reality supports the phenomenal world. In the next chapter I examine some of the contradictory quantum realities which different physicists claim to be the "real reality" that lies behind the external appearances of this world we live in.

2

Physicists
Losing Their Grip

No development of modern science has had a more profound impact on human thinking than the advent of quantum theory. Wrenched out of centuries-old thought patterns, physicists of a generation ago found themselves compelled to embrace a new metaphysics. The distress which this reorientation caused continues to the present day. Basically physicists have suffered a severe loss: their hold on reality.

Bryce DeWitt
Neill Graham

One of the best-kept secrets of science is that physicists have lost their grip on reality.

News of the reality crisis hardly exists outside the physics community. What shuts out the public is partly a language barrier—the mathematical formalism that facilitates communication between scientists is incomprehensible to outsiders—and partly the human tendency of physicists to publicize their successes while soft-pedalling their confusions and uncertainties. Even among themselves, physicists prefer to pass over the uncomfortable reality issue in favor of questions "more concrete". Recent popularizations such as Heinz Pagels' *Cosmic Code* have begun to inform the public about the reality crisis in physics. In *Quantum Reality* I intend to examine how physicists deal with reality—or fail to deal with it—in clear and unprecedented detail.

Nothing exposes the perplexity at the heart of physics more starkly than certain preposterous-sounding claims a few outspoken physicists are making concerning how the world really works. If we take these claims at face value, the stories physicists tell resemble the tales of mystics and madmen. Physicists are quick to reject such unsavory associations and insist that they speak sober fact. We do not make these claims out of ignorance, they say, like ancient mapmakers filling in terra incognitas with plausible geography. Not ignorance, but the emergence of unexpected knowledge forces on us all new visions of the way things really are.

The new physics vision is still clouded, as evidenced by the multiplicity of its claims, but whatever the outcome it is sure to be far from ordinary. To give you a taste of quantum reality, I summarize here the views of its foremost creators in the form of eight realities which represent eight major guesses as to what's really going on behind the scenes. Later we will look at each of these realities in more detail and see how different physicists use the same data to justify so many different pictures of the world.

Quantum Reality #1: The Copenhagen interpretation, Part I (There is no deep reality.) No one has influenced more our notions of what the quantum world is really about than Danish physicist Niels Bohr, and it is Bohr who puts forth one of quantum physics' most outrageous claims: that there is no deep reality. Bohr does not deny the evidence of his senses. The world we see around us is real enough, he affirms, but it floats on a world that is not as real. Everyday phenomena are themselves built not out of phenomena but out of an utterly different kind of being.

Far from being a crank or minority position, "There is no deep reality" represents the prevailing doctrine of establishment physics. Because this quantum reality was developed at Niels Bohr's Copenhagen institute, it is called the "Copenhagen interpretation." Undaunted by occasional challenges by mavericks of realist persuasion, the majority of physicists swear at least nominal allegiance to Bohr's anti-realist creed. What more glaring indication of the depth of the reality crisis than the official rejection of reality itself by the bulk of the physics community?

Einstein and other prominent physicists felt that Bohr went too far in his call for ruthless renunciation of deep reality. Surely all Bohr meant to say was that we must all be good pragmatists and not extend our speculations beyond the range of our experiments. From the results of experiments carried out in the twenties, how could Bohr conclude that no future technology would ever reveal a deeper truth? Certainly Bohr never intended actually to *deny* deep reality but merely counseled a cautious skepticism toward speculative hidden realities.

Bohr refused to accept such a watered-down version of the Copenhagen doctrine. In words that must chill every realist's heart, Bohr insisted: *"There is no quantum world.* There is only an abstract quantum description."

Werner Heisenberg, the Christopher Columbus of quantum theory, first to set foot on the new mathematical world, took an equally tough stand against reality-nostalgic physicists such as Einstein when he wrote: "The hope that new experiments will lead us back to objective events in time and space is about as well founded as the hope of discovering the end of the world in the unexplored regions of the Antarctic."

The writings of Bohr and Heisenberg have been criticized as obscure and open to many interpretations. Recently Cornell physicist N. David Mermin neatly summed up Bohr's anti-realist position in words that leave little room for misunderstanding: "We now know that the moon is de monstrably not there when nobody looks." (We will take a look at Mermin's "demonstration" in Chapter 13.)

Quantum Reality #2: The Copenhagen interpretation, Part II (Reality is created by observation.) Although the numerous physicists of the Copenhagen school do not believe in deep reality, they do assert the existence of *phenomenal reality.* What we see is undoubtedly real, they say, but these phenomena are not really there in the absence of an observation. The Copenhagen interpretation properly consists of two distinct parts: 1. There is no reality in the absence of observation; 2. Observation creates reality. "You create your own reality," is the theme of Fred Wolf's *Taking the Quantum Leap.*

Which of the world's myriad processes qualify as observations? What special feature of an observation endows it with the power to create reality? Questions like these split the observer-created reality school into several camps, but all generally subscribe to quantum theorist John Wheeler's memorable maxim for separating what is real in the world from what is not. "No elementary phenomenon is a real phenomenon until it is an observed phenomenon," Wheeler proclaims. Without a doubt, Mermin's description of the inconstant moon qualifies him for membership in the observer-created reality school.

The belief that reality is observer-created is commonplace in philosophy, where it serves as the theme for various forms of idealism. Bertrand Russell recalls his fascination with idealism during his student days at Trinity College: "In this philosophy I found comfort for a time . . . There was a curious pleasure in making oneself believe that time and space are unreal, that matter is an illusion and that the world really consists of nothing but mind."

Since pondering matter is their bread and butter, not many physicists would share Russell's enjoyment of matter as mere mirage. However, like it or not, through their conscientious practice of quantum theory more than a few physicists have strayed within hailing distance of the idealist's dreamworld.

Quantum Reality #3 (Reality is an undivided wholeness.) The views of Walter Heitler, author of a standard textbook on the light/matter interaction, exemplify a third unusual claim of quantum physicists: that in spite of its obvious partitions and boundaries, the world in actuality is a seamless and inseparable whole—a conclusion which Fritjof Capra develops in *Tao of Physics* and connects with the teachings of certain oriental mystics. Heitler accepts an observer-created reality but adds that the act of observation also dissolves the boundary between observer and observed: "The observer appears, as a necessary part of the whole structure, and in his full capacity as a conscious being. The separation of the world into an 'objective outside reality' and 'us,' the self-conscious onlookers, can no longer be maintained. Object and subject have become inseparable from each other."

Physicist David Bohm of London's Birkbeck College has especially stressed the necessary wholeness of the quantum world: "One is led to a new notion of unbroken wholeness which denies the classical analyzability of the world into separately and independently existing parts . . . The inseparable quantum interconnectedness of the whole universe is the fundamental reality."

Quantum wholeness is no mere replay of the old saw that everything is connected to everything else, no twentieth-century echo, for instance, of Newton's insight that gravity links each particle to every other. All ordinary connections—gravity, for one—inevitably fall off with distance, thus conferring overwhelming importance on nearby connections while distant connections become irrelevant. Undoubtedly we are all connected in unremarkable ways, but close connections carry the most weight. Quantum wholeness, on the other hand, is a fundamentally new kind of togetherness, undiminished by spatial and temporal separation. No casual hookup, this new quantum thing, but a true mingling of distant beings that reaches across the galaxy as forcefully as it reaches across the garden.

Quantum Reality #4: The many-worlds interpretation (Reality consists of a steadily increasing number of parallel universes.) Of all claims of the New Physics none is more outrageous than the contention that myriads of universes are created upon the occasion of each measurement act. For any situation in which several different outcomes are possible (flipping a coin, for instance), some physicists believe that *all outcomes actually occur.* In order to accommodate different outcomes without contradiction, entire new universes spring into being, identical in every detail except for the single outcome that gave them birth. In the case of a flipped coin, one universe contains a coin that came up heads; another, a coin showing tails. Paul Davies champions this claim, known as the many-worlds interpretation, in his book *Other Worlds.* Science fiction writers commonly invent parallel universes for the sake of a story. Now quantum theory gives us good reason to take such stories seriously.

Writing in *Physics Today,* a major magazine of the American physics community, Bryce DeWitt describes his initial contact with the many-worlds interpretation of quantum theory:

"I still recall vividly the shock I experienced on first encountering this multiworld concept. The idea of $10^{100}+$ slightly imperfect copies of oneself all constantly splitting into further copies, which ultimately become unrecognizable, is not easy to reconcile with common sense . . ."

Invented in 1957 by Hugh Everett, a Princeton graduate student, the many-worlds interpretation is a latecomer to the New Physics scene. Despite its bizarre conclusion, that innumerable parallel universes each as real as our own actually exist, Everett's many-worlds picture has gained considerable support among quantum theorists. Everett's proposal is particularly attractive to theorists because it resolves, as we shall see, the major unsolved puzzle in quantum theory—the notorious quantum measurement problem.

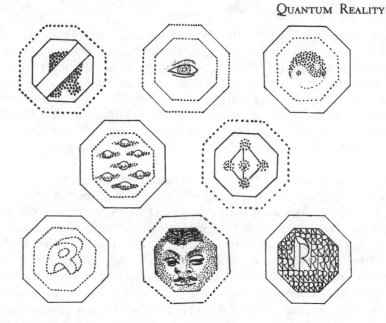

FIG. 2.1   *Eight emblems of quantum reality: which of these worlds, if any, lies behind the quantum facts?*

These four quantum realities should give you some feeling for the diversity of claims regarding the world's ultimate nature. While followers of Everett bear witness to uncountable numbers of quantum worlds, plus more on the way, students of Bohr and Heisenberg insist that there is *not even one* quantum world. In their struggle to gain firm footing amidst the slippery bricks of quantum fact, physicists have invented more realities than four. Keep your wits about you as we press on.

Quantum Reality #5: Quantum logic (The world obeys a non-human kind of reasoning.). Quantum logicians argue that the quantum revolution goes so deep that replacing new concepts with old will not suffice. To cope with the quantum facts we must scrap our very mode of reasoning, in favor of a new quantum logic.

Logic is the skeleton of our body of knowledge. Logic spells out how we use some of the shortest words in the language, words such as *and, or,* and *not.* The behavior of these little linguistic connectors governs the way we talk about things, and structures, in turn, the way we think about them.

For two thousand years, talk about logic (in the West) was cast in the syllogistic mold devised by Aristotle. In the mid-nineteenth century, George Boole, an Irish schoolteacher, reduced logical statements to simple arithmetic by inventing an artificial symbolic language which laid bare the logical bones of ordinary language.

Boole's clear codification of the rules of reason jolted logic out of the Middle Ages and launched the now-flourishing science of mathematical logic. Outside the mathematical mainstream, a few creative logicians amused themselves by constructing "crazy logics" using rules other than Boole's. These deviant designs for *and/or/not*, although mathematically consistent, were considered mere curiosities since they seemed to fit no human pattern of discourse.

However, according to some New Physicists, one of these crazy logics may be just what we need to make sense out of quantum events. Listen to quantum theorist David Finkelstein calling for mutiny against the rules of Boole:

"Einstein threw out the classical concept of *time;* Bohr throws out the classical concept of *truth* . . . Our classical ideas of logic are simply wrong in a basic practical way. The next step is to learn to think in the right way, to learn to think quantum-logically."

As an example of the usefulness of changing your mind rather than changing your physics, quantum logicians point to Einstein's general theory of relativity, which achieved in the realm of geometry what they propose to do with logic.

Geometry is the science of points and lines. For two thousand years only one geometry existed, its rules compiled by the Greek mathematician Euclid in his bestselling book *The Elements*, which once rivaled the Bible in popularity. The latest revival of Euclid's *Elements* is your high school geometry book.

Coincident with Boole's pioneer work in logic, a few adventurous mathematicians thought up "crazy geometries," games points and lines could play outside of Euclid's rules. Chief architect of the New Geometry was the Russian Nicolai Lobachevski along with German mathematicians Karl Gauss and Georg Riemann. Their cockeyed geometries were regarded, like non-Boolean logics, as high mathematical play, clever business but out of touch with reality. Euclidean geometry, as everyone knows, was *the* geometry, being after all, nothing but common sense applied to triangles and other geometric figures.

However, in 1916 Einstein proposed a radical new theory of gravity that demolished the Euclidean monopoly. Einstein, in opposition to Newton and everybody else, declared that *gravity is not a force but a curvature in space-time.* Objects in free fall are truly free and move in lines as straight as can be—that is, lines straight by the standards of a *gravity-warped geometry.* Einstein's theory has testable consequences: for instance the deflection of starlight grazing the sun (confirmed by Eddington in 1919) and the existence of black holes (according to astrophysicists, in the constellation Cygnus, black hole Cygnus X-1 resides). On Earth, where our common sense was formed, gravity is weak and space almost Euclidean; out near X-1, high school geometry flunks.

Einstein's lesson is plain to see, say the quantum logicians. The question of the world's true geometry is not settled by common sense but by experiment. Likewise with logic. For the rules of right reason, look not inside your own head but get thee to a laboratory.

Quantum Reality #6: Neorealism (The world is made of ordinary objects.) An *ordinary object* is an entity which possesses attributes of its own whether observed or not. With certain exceptions (mirages, illusions, hallucinations), the world outside seems populated with objectlike entities. The clarity and ubiquity of ordinary reality has seduced a few physicists—I call them neorealists—into imagining that this familiar kind of reality can be extended into the atomic realm and beyond. However, the unremarkable and common-sense view that ordinary objects are themselves made of objects is actually the blackest heresy of establishment physics.

"Atoms are not things," says Heisenberg, one of the high priests of the orthodox quantum faith, who likened neorealists to believers in a flat earth. "There is no quantum world," warned Bohr, the pope in Copenhagen; "there is only an abstract quantum description."

Neorealists, on the other hand, accuse the orthodox majority of wallowing in empty formalism and obscuring the world's simplicity with needless mystification. Instead they preach return to a pure and more primitive faith. Chief among neorealist rebels was Einstein, whose passion for realism pitted him squarely against the quantum orthodoxy: "The Heisenberg-Bohr tranquilizing philosophy—or religion?—is so delicately contrived that, for the time being, it provides a gentle pillow for the true believer from which he cannot very easily be aroused. So let him lie there."

Despite their Neanderthal notions, no one could accuse neorealists of ignorance concerning the principles of quantum theory. Many of them were its founding fathers. Besides Einstein, prominent neorealists include Max Planck, whose discovery of the constant of action sparked the quantum revolution; Erwin Schrödinger, who devised the wave equation every quantum system must obey; and Prince Louis de Broglie, who took quantum theory seriously enough to predict the wave nature of matter.

De Broglie, a French aristocrat whose wartime involvement in radio swerved his research from church history into physics, fought for ordinary realism until 1928 when he converted to the *statistical interpretation* (another name for Copenhagenism). Twenty years later, however, influenced by David Bohm's neorealist revival, de Broglie recanted and returned to the faith of his youth:

"Those interested in the psychology of scientists may be curious about the reasons for my unexpected return to discarded ideas . . . I am thinking not so much of my constant difficulties in developing a statistical interpretation of wave mechanics, or even of my secret hankering after Cartesian clarity in the midst of the fog which seemed to envelop quantum physics [but the fact that, as I examined the statistical picture] I could not help being struck by the force of the objections to it and by a certain obscurity in the arguments in its defense . . . too abstract . . . too schematic . . . I realized that I had been seduced by the current fashion, and began to understand why I had been so uneasy whenever I tried to give a lucid account of the probability interpretation."

One of the physics community's few traditions is the custom of celebrating the birthdays of its great men with a *Festschrift*—a festival of papers. In 1982, Louis de Broglie, ninety years old and gloriously unrepentant, was honored in this scholarly manner by his scientific colleagues. Virtually every neorealist in the world attended de Broglie's birthday party: there was no need to send out for extra chairs.

Einstein, despite his numerous contributions to its success, never accepted quantum theory into his heart and stubbornly held to the old-fashioned belief that a realistic vision of the world was compatible with the quantum facts. During the thirties Einstein and Bohr engaged in an extended debate on the quantum reality question. Bohr argued that as far as reality was concerned, quantum theory was a closed book. By 1928 perceptive physicists had already grasped the theory's essence. Quantum theory would develop in detail but its principles would not change. Bohr's confidence has been upheld so far; fifty years later, physicists still follow the old rules.

Quantum theory is complete as it stands, said Bohr. It has no need of ordinary objects. Furthermore such objects cannot be added without spoiling its predictive success. Ordinary objects are not merely unnecessary luxuries in quantum theory, they are strictly impossible.

Einstein's strategy was to confront Bohr with a series of thought experiments which aimed to show that quantum theory had left something out. He did not attempt to show that the theory was *wrong*, but by demonstrating that it was *incomplete* Einstein hoped to open the door for what he called "elements of reality."

As the winners tell the story, Bohr closed each of Einstein's loopholes, but in the minds of each the debate was never settled. Long after their arguments had ended, on the day Bohr died, his blackboard contained a drawing of one of Einstein's thought experiments. Bohr struggled with Einstein to the end.

Einstein too never gave up. In his autobiography he expresses his final thoughts on the quantum reality question: "I still believe in the possibility of a model of reality—that is, of a theory which represents things themselves and not merely the probability of their occurrence."

Quantum Reality #7 (Consciousness creates reality.) Among observer-created realists, a small faction asserts that only an apparatus endowed with consciousness (even as you and I) is privileged to create reality. The one observer that counts is a conscious observer. Denis Postle examines reality-creating consciousness in *Fabric of the Universe.* I include this quantum reality not only because it is so outlandish but because its supporters are so illustrious. Consciousness-created reality adherents include light/matter physicist Walter Heitler, already cited in connection with undivided wholeness, Fritz London, famous for his work on quantum liquids, Berkeley S-matrix theorist Henry Pierce Stapp, Nobel laureate Eugene Wigner, and world-class mathematician John von Neumann.

Hungarian-born von Neumann was the mathematical midwife for some of the twentieth century's most exciting developments. Wherever things were hottest, the brilliant von Neumann seemed to be there lending a hand. In the late forties he invented the concept of the stored-program computer; today's computer scientists refer to all computers from pocket calculators to giant IBMs as "von Neumann machines." In collaboration with Oskar Morgenstern, von Neumann laid the mathematical foundation for strategic game theory, on which much government and corporate policy in both the East and the West is based. He also worked on early robots and helped develop the atom bomb. In 1936 with Harvard mathematician Garrett Birkhoff he came up with the idea of quantum logic, but von Neumann's biggest contribution to quantum reality research was his book on quantum theory.

By the late twenties physicists had constructed a quantum theory that met their daily needs: they possessed a rough mathematical structure which organized the quantum facts. At that point von Neumann entered the picture, putting physicists' crude theory into rigorous form, settling quantum theory into an elegant mathematical home called "Hilbert space" where it resides to this day, and awarding the mathematician's seal of approval to physicists' fledgling theory.

In 1932 von Neumann set down his definitive vision of quantum theory in a formidable tome entitled *Die Mathematische Grundlagen der Quantenmechanik*. Our most general picture of quantum theory is essentially the same as that outlined by von Neumann in *Die Grundlagen* (The Foundations). Von Neumann's book is our quantum bible. Like many other sacred texts, it is read by few, venerated by many. Despite its importance it was not translated into English until 1955.

Many of the issues I discuss in *Quantum Reality* were first made public in von Neumann's book. For instance, there is *von Neumann's proof* that if quantum theory is correct, the world cannot be made of ordinary objects —i.e., the neorealist interpretation is logically impossible. Von Neumann posed, but did not solve to everyone's satisfaction, the famous quantum measurement problem which is the central issue of the quantum reality question. In addition, von Neumann was the first to show how quantum theory suggests an active role for the observer's consciousness. Physical objects would have no attributes, von Neumann said, if a conscious observer were not watching them.

Von Neumann himself merely hinted at consciousness-created reality in dark parables. His followers boldly took his arguments to their logical conclusion: if we accept von Neumann's version of quantum theory, they say, a consciousness-created reality is the inevitable outcome.

At the logical core of our most materialistic science we meet not dead matter but our own lively selves. Eugene Wigner, von Neumann's Princeton colleague and fellow Hungarian (they went to the same high school in Budapest), comments on this ironic turn of events: "It is not possible to formulate the laws of quantum mechanics in a fully consistent way without reference to the consciousness . . . It will remain remarkable in whatever way our future concepts may develop, that the very study of the external world led to the conclusion that the content of the consciousness is an ultimate reality."

Quantum Reality #8: The duplex world of Werner Heisenberg (The world is twofold, consisting of potentials and actualities.) Most physicists believe in the Copenhagen interpretation, which states that there is no deep reality (QR #1) and observation creates reality (QR #2). What these two realities have in common is the assertion that *only phenomena are real;* the world beneath phenomena is not.

One question which this position immediately brings to mind is this: "If observation creates reality, what does it create this reality out of? Are phenomena created out of sheer nothingness or out of some more substantial stuff?" Since the nature of unmeasured reality is unobservable by definition, many physicists dismiss such questions as meaningless on pragmatic grounds.

However, since it describes measured reality with perfect exactness, quantum theory must contain some clues concerning the raw material out of which phenomena spring. Perhaps using the power of imagination we can peer beneath this theory and make some shrewd guess about the background world against which our familiar world of solid observations stands.

Werner Heisenberg was fully aware of the difficulties of attempting to describe the subphenomenal world: "The problems of language here are really serious," he said. "We wish to speak in some way about the structure of the atoms and not only about the 'facts'—for instance, the water droplets in a cloud chamber. But we cannot speak about the atoms in ordinary language." Although he realized the difficulty in doing so, Heisenberg was one of the few physicists to try to express what he saw when he looked into quantum reality.

According to Heisenberg, there is no deep reality—nothing down there that's real in the same sense as the phenomenal facts are real. The unmeasured world is merely semireal, and achieves full reality status during the act of observation: "In the experiments about atomic events we have to do with things and facts, with phenomena that are just as real as any phenomena in daily life. But the atoms and the elementary particles themselves are not as real; they form a world of potentialities or possibilities rather than one of things or facts . . .

"The probability wave . . . means a tendency for something. It's a quantitative version of the old concept of *potentia* in Aristotle's philosophy. It introduces something standing in the middle between the idea of an event and the actual event, a strange kind of physical reality just in the middle between possibility and reality."

Heisenberg's world of potentia is both less real and more real than our own. It is less real because its inhabitants enjoy a ghostly quantum lifestyle consisting of mere tendencies, not actualities. On the other hand, the unmeasured world is more real because it contains a wealth of coexistent possibilities, most of which are contradictory. In Heisenberg's world a flipped coin can show heads and tails at the same time, an eventuality impossible in the actual world.

One of the inevitable facts of life is that all of our choices are real choices. Taking one path means forsaking all others. Ordinary human experience does not encompass simultaneous contradictory events or multiple histories. For us, the world possesses a singularity and concreteness apparently absent in the atomic realm. Only one event at a time happens here; but that event really happens.

The quantum world, on the other hand, is not a world of actual events like our own but a world full of numerous unrealized tendencies for action. These tendencies are continually on the move, growing, merging, and dying according to exact laws of motion discovered by Schrödinger and his colleagues. But despite all this activity *nothing ever actually happens there*. Everything remains strictly in the realm of possibility.

Heisenberg's two worlds are bridged by a special interaction which physicists call a "measurement." During the magic measurement act, one quantum possibility is singled out, abandons its shadowy sisters, and surfaces in our ordinary world as an actual event. Everything that happens in our world arises out of possibilities prepared for in that other—the world of quantum potentia. In turn, our world sets limits on how far crowds of potentia can roam. Because certain facts are actual, not everything is possible in the quantum world. There is no deep reality, no deep reality-as-we-know-it. Instead the unobserved universe consists of possibilities, tendencies, urges. The foundation of our everyday world, according to Heisenberg, is no more substantial than a promise.

Physicists do not put forth these quantum realities as science fiction speculations concerning worlds that might have been, but as serious pictures of the one world we actually live in: the universe outside your door.

Since these quantum realities differ so radically, one might expect them to have radically different experimental consequences. An astonishing feature of these eight quantum realities, however, is that they are experimentally indistinguishable. For all presently conceivable experiments, *each of these realities predicts exactly the same observable phenomena.*

The ancient philosophers faced a similar reality crisis. For instance three ancient realities—1. The world rests on a turtle's back; 2. The world is bottomlessly solid; 3. The world floats in an infinite ocean—led to identical consequences as far as anyone could tell at that time.

Likewise modern physicists do not know how to determine experimentally what kind of world they actually live in. However, since "reality has consequences" we might hope that future experiments, not bound by our current concepts of measurability, will conclusively establish one or more of these bizarre pictures as top-dog reality. At present, however, each of these quantum realities must be regarded as a viable candidate for "the way the world really is." They may, however, all be wrong.

Physicists' reality crisis is twofold: 1. There are too many of these quantum realities; 2. All of them without exception are preposterous. Some of these quantum realities are compatible with one another. For instance QR #1 (There is no deep reality) and QR #2 (Reality is observer-created) are in fact two halves of a single consistent picture of the world called the Copenhagen interpretation. But other quantum realities are contradictory: in the many-worlds interpretation (QR #4), for instance, the world's deep reality consists of quadrillions of simultaneous universes, each one as real as our own, which maximally mocks Bohr's no-deep-reality claim. Not only can physicists not agree on a single picture of what's really going on in the quantum world, they are not even sure that the correct picture is on this list.

None of the conflicting options which physicists have proposed as possible pictures of our home universe can be considered ordinary. Even that quantum reality closest to old-fashioned notions of how a world should behave —the neorealist contention (QR #6) that the world is made out of ordinary objects—contains, as we shall see, the requirement that some of these objects move faster than light, a feature that entails unusual consequences: time travel and reversed causality, for example.

This book is a snapshot of the reality crisis in physics taken at a moment when that crisis is not yet resolved. Nobody knows how the world will seem one hundred years from now. It will probably appear very different from what we now imagine. Here's what John Wheeler, a physicist actively concerned with the nature of quantum reality, imagines when he looks into the future:

"There may be no such thing as the 'glittering central mechanism of the universe' to be seen behind a glass wall at the end of the trail. Not machinery but magic may be the better description of the treasure that is waiting."

3

Quantum Theory
Takes Charge

Some physicists would prefer to come back to the idea of an objective real world whose smallest parts exist objectively in the same sense as stones or trees exist independently of whether we observe them. This however is impossible.

Werner Heisenberg

At the end of the nineteenth century, physicists possessed a comprehensive picture of the way the world worked. A few great men had solved the big problems. The task of their successors was to fill in the details, to measure the next decimal place. No glory there. By explaining everything, classical physics seemed to have put itself out of business. "Physics is finished, young man. It's a dead-end street," said Max Planck's teacher. He advised Planck to be a concert pianist instead.

The triumph of classical physics was short-lived: the paradoxes of quantum theory soon swept away its Victorian certainties. But while it lasted, nineteenth-century physics stood as a high-water mark of applied common sense. Not only did it appear to explain all the facts, but it did so in ways that were clear and picturable.

MATTER AND FIELD—THE STUFF OF CLASSICAL PHYSICS

Classical physicists were able to account for all the world's variety by means of only two physical entities—matter and fields. In those innocent days it went without saying that these entities were *really there*. Physicists' reality crisis was yet to come.

Real matter. Real fields. Drop an apple from a bridge. The apple is made of matter. It moves because the Earth's gravitational field pulls it. Everything in the world works the same way: matter produces force fields, which move other matter.

Classical physics recognized just two fields—electromagnetic and gravitational—leaving open the possibility that more might be discovered. Modern physics added only two fields to the classical duo—the *strong field*, which holds the atomic nucleus together, and the *weak field*, which breaks the nucleus apart in certain kinds of radioactive decay. Present wisdom holds that these are all the fields nature needs to produce our kind of universe, and that they are all likely one, united like electricity and magnetism.

A classical field is a distribution of forces in space. The range of a field is how far its force extends. The three classical forces fall off as the distance squared, but they never go to zero: the range of these classical forces is infinite. No matter how far from Earth you travel, its gravity still pulls on you a little. On the other hand, the force of the two modern fields is confined to the atomic nucleus. Because of their short ranges, the weak and strong force fields were discovered last.

To describe how the classical world works we need two kind of laws: laws of motion and field laws. Laws of motion tell matter how to move in a particular force field. Field laws describe how each field depends on its material source and how it spreads itself through space.

In the seventeenth century Isaac Newton discovered matter's laws of motion. Force fields push matter along paths exactly prescribed by Newton's laws. These laws are *deterministic:* a given situation always leads to a unique outcome. Locked to its tracks, the world cannot help but follow a single path. Second by second the universe, like a giant clockwork, ticks out Newton's inexorable laws, its future as fixed and immutable as its past.

In addition to his laws of motion, Newton discovered physics' first field law—the inverse square behavior of the gravity field. To complete the classical picture, only the field laws of electricity and of magnetism were lacking. During the U.S. Civil War, Scottish physicist James Clerk Maxwell closed this gap, laying down the laws that govern electric and magnetic fields. Maxwell's laws were full of surprises. For instance, electricity and magnetism turned out to be not two separate fields but different aspects of a single electromagnetic field. Maxwell's field laws also unexpectedly solved one of physics' long-standing mysteries: the intrinsic nature of light.

Fields whose range extends to the distant stars are attached to each piece of matter. Shake that matter and you shake its field. Motion in a field, like motion in a water bed, doesn't stand still but wiggles away as fast as it can. Shaking a field makes waves—waves that travel at a certain velocity.

Maxwell's laws tell us how to calculate the speed of waves in the electromagnetic field. How fast these waves move depends entirely on two electromagnetic facts: the force between two magnets and the force between two electric charges. From the measured magnitudes of magnetic and electric attraction, Maxwell figured how fast a wave of electromagnetism must travel. His calculated speed was identical to the measured speed of light. Maxwell conjectured that light was actually an electromagnetic wave of extremely high frequency. Prior to Maxwell, who would have guessed that this tenuous radiance that fills our eyes is akin to stolid storage batteries and industrial magnets?

Furthermore, Maxwell surmised that there must exist invisible electromagnetic waves both lower and higher than light in the electromagnetic spectrum. Heinrich Hertz's subsequent production of radio waves ("low-frequency light") and the discovery of X rays ("high-frequency light") by Roentgen verified Maxwell's bold conjecture: Light is a wave motion in the electromagnetic field. Maxwell's discovery that the colors which delight the eye, the needle that guides the sailor, the lodestone, lightning, the electric twitch in muscle and brain, are all manifestations of a single physical field was classical physics' finest hour.

Ripples presumably exist in the gravity field, but gravity waves are too weak to influence even the most delicate gravity meter. One of modern physics' frontiers is the development of instruments sensitive enough to respond to waves of gravity.

Classical physics in a nutshell: The universe consists of nothing but matter and fields—and we know the laws of both. What more could a physicist wish for? Well, it seems the picture was not quite perfect. If you pressed them, classical physicists would confess a few tiny blemishes which they were sure could be erased with a little extra effort. There was, for instance, the black-body radiation puzzle.

WHY DO HOT OBJECTS GLOW RED?

Colored objects have an intrinsic color; black objects don't. Heat up a black object, however, and it begins to glow. Steelmakers for generations gauged furnace temperatures by this black-body glow. They know, for instance, that iron turns cherry-red around thirteen hundred degrees. For physicists the black-body puzzle is how to calculate the color of that glow at different temperatures.

A black object is made of little pieces of matter. Whenever these pieces move, they shake waves into their attached electromagnetic fields—waves our eyes interpret as colored light. The faster the particles move, the higher the frequency of the light that is shaken off. As an object gets hotter, its parts move faster. That's the sort of thinking that goes on in a classical physicist's head as he sets out to calculate the color of black-body glow.

Classical physicists had little idea of the nature of the light-emitting particles in a block of hot iron but they assumed that, like everything else in the world, they obeyed Newton's laws. Today we know that light is caused by moving electrons. However, not only do electrons not follow classical laws, they do not even follow a classical *kind* of law—that is, a law that governs the motion of real objects.

For a quarter century after Maxwell announced the light-matter connection, physicists attacked the black-body puzzle and kept coming up with the same answer: black bodies should glow bright blue at all temperatures.

In 1900, as the new century began, Max Planck, who against his teacher's advice had earned a degree in physics, not music, took up this black-body puzzle. As a simplifying assumption he decided not to let the matter particles vibrate any way they pleased; instead he artificially constrained them to frequencies that follow this simple rule:

$$E = nhf$$

where E is the particle's energy, n is any integer, f is the frequency of the particle's vibration, and h is a constant to be chosen by Planck. Planck's rule restricts the particles to energies that are certain multiples of their vibration frequency, as though energy only came in "coins" of denomination hf. Planck's constant h would later be called the "quantum of action" because it has the dimensions of energy times time, a quantity known as "action" in classical physics.

Planck's assumption was not justified by any physical reasoning but was merely a trick to make the math easier to handle. Later in his calculations Planck planned to remove this restriction by letting the constant h go to zero. This would make the value of the "energy coin" so small that the particle could once again have effectively any energy it pleased.

Planck discovered that he got the same blue glow as everybody else when h went to zero. However, much to his surprise, if he set h to *one particular value*, his calculation matched the experiment exactly (and vindicated the experience of ironworkers everywhere). Hot iron glows red, Planck showed, only if those particles exist whose energy is built from "coins" of a particular denomination. Physicists politely ignored Planck's work because although it gave the right answer, it did not play fair. This funny restriction on energy was totally alien to classical physics. Newton's laws permitted particles to have any energy they pleased.

EINSTEIN'S THREE PAPERS

In the year 1905 Albert Einstein, a German Jew working in Switzerland as an obscure patent clerk, published three papers in the German journal *Annalen der Physik* that exploded onto the physics scene. Each of these papers was a bombshell that shook the foundations of physics. Each by itself would have sufficed to establish Einstein as a physicist of highest rank. Three at once suggests divine inspiration.

Einstein's first paper explained the photoelectric effect—light's ability to knock electrons out of metal—using Max Planck's new quantum of action. Einstein's analysis demonstrated the non-classical nature of light so unmistakably that physicists could no longer ignore the mysterious quantum. Planck kicked off this new quantum ball game; Einstein made the first touchdown.

Einstein's second paper explained the Brownian motion of microscopic particles in liquids and showed how the centuries-old question of the reality of atoms could be decided experimentally. French physicist Jean Perrin carried out these experiments and established for the first time conclusive evidence for the existence of atoms.

His first paper swept physicists headlong into the Quantum Era; his second decisively settled one of the nineteenth century's biggest controversies. But Einstein was just warming up.

Einstein's third paper profoundly changed our ordinary ideas of time and space, notions that seemed so deeply embedded in human experience as to be unquestionable. In Einstein's new vision, measurements of length and time are not absolute but depend on the observer's velocity. For example, two people watching the same clock see it running at different rates if they are moving relative to each other. Other absolutes take the place of space and time, notably the speed of light, which Einstein declared to be the same for all observers. The speed of light is also elevated to a universal speed limit which no signal can exceed. Einstein's special theory of relativity (as he called it) had profound consequences for physics and philosophy, for it showed that some of our most cherished notions about the world are simply wrong and must be replaced with entirely new ways of thinking.

Because it overturned common-sense ideas, Einstein's relativity is often considered part of the New Physics but strictly speaking relativity belongs to the nineteenth century. Despite its bizarre notions of space and time, Einstein's theory does not challenge classical physics but completes it.

Classical physics is based on two sets of laws: Maxwell's field laws and Newton's motion laws. These two sets of laws could not both be reconciled with the newly observed fact that despite the Earth's motion the velocity of light was constant in all directions. Something in classical physics had to give. Einstein kept Maxwell's laws intact but replaced Newton's laws with his own relativistic laws of motion. These new laws make light's speed a constant for all observers. One consequence of Einstein's revision of Newton is his famous $E = mc^2$ relation. Maxwell's field laws plus the relativistic laws of motion now completely and consistently describe all classical motion even at high velocity.

Because relativity is really part of classical physics, most physicists, once they recovered from their initial shock, learned to accept it as a natural extension of common sense. Despite his radical revision of space and time, Einstein's attitude toward reality is no different from Newton's. These are real fields here; real matter we are dealing with. Special relativity caused no reality crisis in physics.

It is ironic that in the same year Einstein perfected the classical picture of the world, he also began a line of thought (Einstein's first paper) which would utterly destroy it. By his explanation of the photoelectric effect in terms of light quanta, Einstein attacked classical physics not at some obscure point on its fringes but squarely on center. Einstein challenged physicists' understanding of light, a question believed to have been settled by Maxwell before Einstein was born.

THE PHOTOELECTRIC EFFECT:
HOW DOES LIGHT INTERACT WITH ELECTRONS IN A METAL?

Consider a beam of light shining on a piece of metal. Classical physicists imagined electrons in a metal as suspended in the atom's electromagnetic field like swimmers floating in quiet waters. Light is a wave in this electromagnetic ocean and can knock electrons out of metal as ocean waves can wash swimmers onto beaches. The bigger the wave, the harder the electron/swimmer is thrown out of the water.

However, photoelectric experiments don't respect this simple analogy. For light of a given frequency, the ejected electron's energy is always the same for the weakest light as well as for the strongest beam. When the beam is intense, more electrons come out but they all have the same energy. This behavior would appear very strange if it happened at the beach: a ripple and a tidal wave (of the same frequency) would throw you onto the sand with the same force. The tidal wave would just throw more people out.

If you want light to give more of its energy to the ejected electron, increasing the light's intensity is not the way to do it. Instead you must increase its frequency. Light's energy evidently depends on its color, not its intensity. Blue light (high-frequency) kicks electrons harder than red light (low-frequency). Experiments say: the higher the frequency, the greater the impact. At the beach this means that fast little ripples near the water's edge are more dangerous—can give you a bigger jolt—than giant low-frequency tidal waves.

Einstein explained these strange facts about light "waves" by a single assumption. Light behaves like a shower of particles, he said, each with energy E given by Planck's expression:

$$E = hf$$

where f is the light's frequency and h is Planck's constant of action—the particular value Planck had to insert to calculate black-body glow correctly.

More than a century of experiments had shown light to be a wave. Maxwell's successful theoretical description even identified what it was— the electromagnetic field—that was making the waves. Einstein nonetheless demonstrated that when light interacts with metals it resembles a shower of particles divided into "coins of energy." Taken together these experiments indicate that, in some manner not easy to visualize, light acts in certain situations as a wave, in others as particles.

Three wave attributes are especially discordant with the particle notion: 1. A wave can spread out over an enormous area, while a particle is confined to a tiny region. 2. A wave is easily split in an infinite variety of ways, some parts going in one direction, some another, while a particle's travel is confined to a single direction. 3. Two waves can interpenetrate like ghosts and emerge unchanged where particles would crash together. Particle and wave seem irreconcilably different, but the nature of light is such that it is able to combine these contradictory attributes in a harmonious way.

THE COMPTON EFFECT:
HOW DOES LIGHT INTERACT WITH ELECTRONS IN A GAS?

Einstein's particles of light (dubbed "photons") showed up again in an experiment performed by American physicist Arthur Holly Compton. In the photoelectric effect, one photon goes in, but none come out: the quantum of light gives all its energy to the electron. Instead of studying light absorption, Compton studied light scattering. By shining an X-ray beam (high-frequency light) into a gas whose electrons were loosely bound, Compton was able to detect both the ejected electron and the recoil photon. His experiment showed that light behaves precisely like a little particle bouncing off the electron, provided that you assign this light particle a momentum p according to the quantum rule:

$$p = hk$$

where k is the light's spatial frequency—the number of wavelengths that fit into one centimeter—and h is Planck's constant of action. In Compton's experiment light acts for all the world like a tiny billiard ball, with momentum and energy given by the magic quantum rules. You can put in a certain frequency of light, do a billiard-ball-style calculation, and predict where to place counters that will catch both the scattered electron and the recoil photon. Photon in the side pocket!

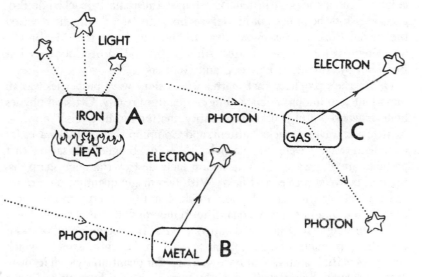

Fɪɢ. 3.1 *Three experiments that made a revolution. A. Black-body radiation; B. the photoelectric effect; C. the Compton effect all indicate that light consists of particles, which conflicts with numerous nineteenth-century optical experiments in which light acts like a wave. Here and elsewhere, elementary quantum entities such as electrons and photons are represented as five-pointed stars*

De Broglie Predicts Wave Nature of Matter

While physicists were puzzling over light's seemingly contradictory properties, another storm was brewing in Paris. French aristocrat Louis de Broglie had submitted a strange Ph.D. thesis to his physics professors at the Sorbonne. By arguments that seemed dubious to his examiners, de Broglie contended that each particle of matter was associated with a wave whose temporal and spatial frequencies f and k were given by the Planck-Einstein recipe $E = hf$ and the Compton relation $p = hk$ where E and p are the particle's energy and momentum. De Broglie argued that just as

Einstein showed waves of light to have particle properties, so particles of matter might also have wave properties.

Here was a delicate situation. Prince de Broglie was serious about his proposal but his conclusion was patently absurd. His thesis professor decided to send a copy to Einstein. Einstein enthusiastically backed de Broglie's idea and the prince got his degree. Six years later de Broglie received the Nobel Prize for his crazy idea. In the meantime the "de Broglie wavelength" of a piece of matter—the electron—had been measured at Bell Labs by Americans Davisson and Germer.

De Broglie's prophecy that matter would show wave properties was an important step in our understanding of quantum reality. Classical physics built its world out of two kinds of entity: matter and field (also known as particle and wave). Planck, Einstein, and Compton showed that waves (at least light waves) were also particles. Now de Broglie was saying that particles are also waves. New quantum facts destroy the once sharp distinction between matter and field. With two magic quantum phrases we can translate at will between the particle quantities energy and momentum $(E$ and $p)$ and the wave quantities temporal and spatial frequency $(f$ and $k)$, turning matter into field and vice versa. It's beginning to look as if everything is made of one substance—call it "quantumstuff"—which combines particle and wave at once in a peculiar quantum style all its own.

By dissolving the matter/field distinction, quantum physicists realized a dream of the ancient Greeks who speculated that beneath its varied appearances the world was ultimately composed of a single substance. Some philosophers said it was All Fire; some All Water. We now believe the world to be All Quantumstuff.

The world is one substance. As satisfying as this discovery may be to philosophers, it is profoundly distressing to physicists as long as they do not understand the nature of that substance. For if quantumstuff is all there is and you don't understand quantumstuff, your ignorance is complete.

French physicist Oliver Costa de Beauregard calls the quarter century from Planck's discovery of the quantum of action (1900) to Heisenberg's formulation of matrix mechanics (1925) the *Era Paléoquantique* or Quantum Stone Age. To explain the quantum facts during these confusing times, physicists pieced together fragments of classical physics with certain quantum notions (notably the two magic phrases connecting wave and particle attributes) in clever but essentially haphazard ways.

The high-water mark of Stone Age techniques was Bohr's model of the hydrogen atom, which explained major details of hydrogen's spectrum—a problem that classical physics could not even touch. Bohr, however, was aware that his success was largely a matter of inspired guesswork. During this transition period physicists lacked a reliable and consistent method—quantum laws comparable to the classical laws of Newton and Maxwell—of dealing with the quantum facts.

Physicists yearned for a quantum theory to lead them out of Stone Age ignorance. The need so intense, the time so ripe, that in a single year not one, not two, but three separate quantum theories arose where before there had been none.

THE BIRTH OF QUANTUM THEORY

Werner Heisenberg was first. In the summer of 1925 Heisenberg was recovering from a hay-fever attack on the North Sea island of Heligoland. Inspired by conversations with Bohr on the quantum mysteries, Heisenberg in his island retreat suddenly put it all together and came up with matrix mechanics—the world's first quantum theory.

Quantum theory is a method of representing quantumstuff mathematically: a model of the world executed in symbols. Whatever the math does on paper, the quantumstuff does in the outside world. Quantum theory must contain at least: 1. some mathematical quantity that stands for quantumstuff; 2. a law that describes how this quantity goes through its changes; 3. a rule of correspondence that tells how to translate the theory's symbols into activities in the world.

Quantum Theory #1: Heisenberg represents a quantum system by a set of matrices, hence the name matrix mechanics. A matrix is a square array of numbers like a mileage table on a road map which lists the distances between various cities. Each Heisenberg matrix represents a different attribute, such as energy or momentum, with the mileage chart's cities replaced by particular values of that attribute. The matrix's diagonal entries represent the probability that the system has that particular attribute value, and the off-diagonal elements represent the strength of non-classical connections between possible values of that attribute. Thus momentum p of an electron is not represented by a number as in classical physics, but by one of these square arrays. Likewise with position x, energy E and any other system attribute: they are all represented by matrices.

The evolution of these matrices follows a particular law of motion which resembles Newton's law in form but contains peculiar differences. One big difference is that unlike numbers, matrices don't commute. This means that the order of matrix multiplication makes a difference. In particular, when p and x are square arrays, p times x is not equal to x times p.

In the winter of 1925, Austrian physicist Erwin Schrödinger and Englishman Paul Dirac independently came up with two more quantum theories.

Quantum Theory #2: Schrödinger represented quantumstuff as a wave form and wrote the quantum laws of motion (Schrödinger's equation) such a wave form must obey. At first Schrödinger believed his waves to be classical waves as real as Maxwell's electromagnetic vibrations, but, as we shall see, the reality status of Schrödinger waves is extremely dubious. Because he represents quantumstuff with wave imagery, Schrödinger's theory is called wave mechanics.

Quantum Theory #3: Dirac symbolized quantumstuff as an arrow (or vector) pointing in a certain direction in an abstract space of many dimensions. Motion of quantumstuff corresponds to rotation of that arrow. To describe how the arrow turns, you must set up some sort of coordinates over the arrow's space (analogous to longitude and latitude lines across the Earth's surface) but, as on Earth, there is a great deal of freedom on how to lay down these imaginary lines. Depending on your choice of coordinates you get, for the same arrow, quantum descriptions that superficially look very different. A big part of Dirac's theory concerns how to change from one coordinate system to another, how to *transform* between seemingly different descriptions of the same rotating arrow. Because of its emphasis on switching descriptions, Dirac called his scheme "transformation theory."

Quantum physics emerged from the Stone Age with an embarrassment of riches—three quantum theories, each claiming to explain the world. As it turned out, all three were right. Dirac was able to show that both Heisenberg's and Schrödinger's theories were special cases of his own rotating-arrow version of quantum theory. Dirac's arrow looks like a cluster of matrices or a wave, depending on what coordinate system you select. Thus despite their different pictures of quantumstuff, all three theories

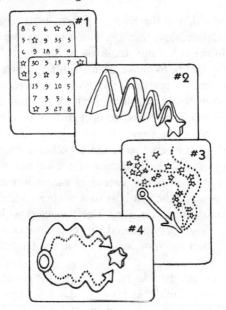

FIG. 3.2 *Four versions of quantum theory in order of discovery: 1. Heisenberg's matrix mechanics; 2. Schrödinger's wave mechanics; 3. Dirac's transformation theory; 4. Feynman's sum-over-histories approach. All four versions give the same results.*

have identical content. Traveling dissimilar paths, Heisenberg, Schrödinger, and Dirac converged on the same explanation from three different directions.

Why stop with three? The freedom to choose coordinates means that quantum theory can describe the same physical situation in a variety of mathematical languages. Physicists exploit this theory's multilingual facility by choosing, for each new problem, whatever language works best. Physicists educated in the prequantum era favored the Schrödinger wave picture because of its kinship with classical physics. However, young physicists weaned on transformation theory soon developed a taste for abstract languages remote from the classical tradition.

When physicists received the triple revelation from Heisenberg, Schrödinger, and Dirac, they began in earnest to test the new theory against events in the outside world. At this point the story of the quantum takes two paths: the tale of those who use quantum theory as a symbolic tool to manipulate the world, and the history of those who regard quantum theory as a window into reality, through which to perceive the world's inner nature. Most physicists became toolmakers; few in the fast-paced quantum era found the leisure to chase after "reality." Much has been written about quantum theory's practical success in every realm from quark to quasar. In *Quantum Reality* I follow the path less traveled.

Quantum Theory was invented to deal with one problem: the interaction of light with atoms. But once the atomic problem was solved, physicists hastened to test the new theory against other mysteries: the nature of solids, liquids, and gases, the structure of the minuscule nucleus hidden deep within the atom, and the nature of subnuclear entities. Some attempted to see beneath the symbols to a deeper reality, but most physicists used quantum theory like a pack of teenagers with a powerful new car, driving it recklessly at top speed till something falls apart. But quantum theory is a tough machine; wherever we want to steer it, this theory takes us there.

Classical physics had its defects, experiments it could not explain, that eventually led to its downfall. As far as we can tell, there is no experiment that quantum theory does not explain, at least in principle. Quantum theory is a perfect match for the quantum facts. Though physicists have steered quantum theory into regions far distant from the atomic realm where it was born, there is no sign on the horizon that it is ever going to break down.

With quantum theory's unprecedented practical success and the powerful control over nature this success delivered, physics became big business, an arm of the state. Reality research was soon eclipsed by an orgy of application.

Physicists were not always so wary of reality. During the first decade of the Quantum Era (1925–35), controversy flourished concerning the quantum theory/reality connection. For instance this period witnessed the famous Bohr-Einstein debates. Gradually the view instigated by Heisenberg, Bohr, and Göttingen-based Max Born prevailed and hardened into the official doctrine known as the Copenhagen interpretation. Copenhagenists claim there is no deep reality (QR # 1).

Some physicists identify Copenhagenism with pragmatism—with what I've called the cookbook approach toward phenomena. Quantum theory is a recipe to be compared with appearances, pragmatists say, and nothing more. Theory tells us nothing about reality, nor should we expect it to do so. Physics is a matter of matching mathematics to measurement; all else is baseless speculation. Pragmatists quote Bohr approvingly: "It is wrong to think that the task of physics is to find out how nature *is*. Physics concerns what we can *say* about nature."

Every theory from horse-race handicap to model of the U.S. economy possesses a minimal pragmatic core: does math fit facts? All theories are pragmatic, at least. Some people are pragmatists because they imagine they can avoid "philosophy" by sticking to the supposed certainties of mathematics and experiments.

In my opinion Bohr was more than a pragmatist. He made definite statements about the absence of a reality beneath quantum theory based not on a distaste for philosophy but using arguments drawn from the particular structure of quantum theory itself. All theories are pragmatic; some theories have a reality underneath. However, quantum theory is not a theory of this type, Bohr contends. As long as it keeps the form discovered by Heisenberg, Schrödinger, and Dirac, quantum theory will never be susceptible of reinterpretation in terms of a deeper reality.

The establishment of the Copenhagen doctrine was an important landmark in reality research because it both represents the majority viewpoint and acted to close off debate on the reality question. As CalTech physicist Murray Gell-Mann puts it: "Niels Bohr brainwashed a whole generation of physicists into thinking that the job was done fifty years ago."

Two other landmarks connected with the Copenhagen interpretation are Heisenberg's uncertainty principle and Bohr's principle of complementarity. Together they express basic restrictions which nature seems to impose on any measurement act. We will see how Bohr uses these fundamental observational limits to argue that neither measurement nor theory can put us in contact with deep reality.

Despite the extreme language of some quantum realists, no physicist actually denies that electrons exist. You see their effect whenever you switch on the TV. If electrons don't exist, what makes the picture on the screen?

What's at stake in the quantum reality question is not the actual existence of electrons but *the manner in which electrons possess their major attributes.* Classical physicists imagined that every particle possessed at each moment a definite position and momentum; each field likewise possessed a particular field strength at every location. If we agree to call any entity—particle, field, apple, or galaxy—which possesses its attributes innately an "ordinary object" then the fundamental message of classical physics was this: the entire physical world consists of nothing but ordinary objects.

Quantum theory suggests, on the other hand, that the world is *not* made of ordinary objects. An electron, and every other quantum entity, does not possess all its attributes innately. An electron does possess certain innate attributes—mass, charge, and spin, for instance—which serve to distinguish it from other kinds of quantum entities. The value of these attributes is the same for every electron under all measurement conditions. With respect to these particular attributes, even the electron behaves like an ordinary object.

However, all other attributes, most notably position and momentum, which, it was thought, classical particles possessed innately, can no longer be attached to the electron without qualification. These attributes—called "dynamic" to distinguish them from the "static" attributes mass, charge, and spin—do not belong to the electron in itself, but seem to be created in part by the electron's measurement context. The manner in which an electron acquires and possesses its *dynamic attributes* is the subject of the quantum reality question. The fact of the matter is that nobody really knows these days how an electron, or any other quantum entity, actually possesses its dynamic attributes.

According to the Copenhagen interpretation, the electron's dynamic attributes are *contextual:* what attributes it seems to have depends on how you measure it. An electron's so-called attributes belong jointly to the electron and the measuring device. When a Copenhagenist says, "There is no deep reality," she means that there is no hidden value of position that the electron "really has" when it is not being measured. Since position is an attribute that belongs jointly to the electron and its measuring device, when you take away the measuring device you take away the electron's position too.

An example of a contextual attribute is the color of an ordinary object. Color is not an innate attribute but depends on the quality of illumination and other aspects of the viewing situation. Grocers exploit the contextual nature of the color attribute by displaying meat and vegetables under tinted lights which render their colors more appealing than the white light of ordinary reality. Although the color of beefsteak is not an innate attribute, it is based on an attribute that is intrinsic to beefsteak—namely, this meat's absorption spectrum. The conjunction of absorption spectrum (innate to beefsteak), emission spectrum (innate to meat-counter light), and spectral response (innate to human eye) determines the color attribute of the steak. Change one element in this triad of innate attributes and you change the beef's perceived color.

Is it possible that the position of an electron is like the color of a beefsteak—that is, a contextual attribute based on deeper attributes which are not contextual? When a Copenhagenist says, "There is no deep reality," she means that the electron is not like a beefsteak: no deep innate attributes exist which explain the electron's measurement-dependent position and momentum. When you take away the measuring device the electron undoubtedly still exists, but it possesses no dynamic attributes at all; in particular it has no definite place or motion. We cannot picture such a state of being, but nature seems to have no trouble producing such entities. Indeed, such entities are all this world is made of.

A beefsteak's color may be contextual, but in most environments it looks red. As long as we keep the limits of such a statement in mind we are entitled to say that beefsteak is red. In other words color is *almost* an innate attribute. But in reality it's not. Likewise, as long as we do not push its limits an electron appears to have a definite position and momentum at all times. The electrons in your TV tube, for instance, when they're not forced to go through tiny holes seem to behave like classical objects. In other words, an electron's dynamic attributes are *almost* innate. The enormous success of classical physics depended on this fact: the entities that make up the world are almost ordinary objects. But in reality they're not.

Von Neumann's Quantum Bible

In 1932 the eminent mathematician John von Neumann published his definitive analysis of quantum theory. Von Neumann's *Die Grundlagen* (The Foundations) is our quantum bible. In this influential book von Neumann gives quantum theory a firm mathematical basis and tackles many reality-related issues in a highly logical manner. Von Neumann poses the famous quantum measurement problem which, as we shall see, lies at the heart of the quantum reality question. It is fair to say that if we could say what actually goes on in a *measurement*, we would know what physical reality was all about. Because of his peculiar views on measurement, von Neumann is sometimes regarded as the godfather of the consciousness-created reality school (QR #7).

In *Die Grundlagen* von Neumann considers the claim of the neorealists that an *ordinary reality* underlies the quantum facts, and in a short formal argument (known to reality researchers as "von Neumann's proof") concludes that the existence of such a reality is mathematically incompatible with quantum theory. Von Neumann's proof was widely regarded as confirming Bohr's Copenhagen party line: there is no deep reality and it is futile to search for one.

What von Neumann showed was that if you assume that electrons are ordinary objects or are constructed of ordinary objects—entities with innate dynamic attributes—then the behavior of these objects must contradict the predictions of quantum theory. Furthermore if you assume that electrons possess contextual attributes that stem from ordinary objects inaccessible to measurement but whose innate attributes combine "in a reasonable way" to simulate the electron's measurement-dependent behavior, then these entities likewise must violate quantum theory's predictions. Thus, according to the quantum bible, *electrons cannot be ordinary objects, nor can they be constructed of (presently unobservable) ordinary objects*. From its mathematical form alone, von Neumann proved that quantum theory is incompatible with the real existence of entities that possess attributes of their own.

For more than a quarter of a century the authority of von Neumann's proof dampened enthusiasm for reality research. Why search for an ordinary reality beneath quantum theory, when the world's top mathematician tells you such a thing is impossible?

However, in 1952, despite von Neumann's proof, David Bohm did the impossible by constructing a model of the electron with innate attributes whose behavior matches the predictions of quantum theory.

BOHM'S ORDINARY-OBJECT MODEL OF THE ELECTRON

Bohm was born in Pennsylvania and studied physics at Berkeley under J. Robert Oppenheimer, who had learned quantum theory in Copenhagen from Bohr himself. When Bohm finished his studies, he became a Princeton professor and decided to clarify his thinking by writing a textbook on quantum theory.

Bohm's textbook *Quantum Theory* is valued by students as a simple introduction in plain English to the mechanics of quantum calculations as well as for its unusually detailed discussions of the reality question. Like his teacher and the majority of physicists, Bohm was a loyal Copenhagenist. His treatment of the reality question follows Bohr's party line. In place of von Neumann's highly mathematical injunction against ordinary reality, Bohm comes to the same conclusion using informal arguments. In *Quantum Theory* Bohm argues, in agreement with Heisenberg, Bohr, and von Neumann, that *electrons are not things.*

In 1951 David Bohm tangled with American political reality when he refused to testify against Oppenheimer before Senator Joseph McCarthy's Committee on Un-American Activities. He lost his job at Princeton and never again taught in the United States, moving first to Brazil and finally settling in London.

At about this time, conversations with Einstein convinced Bohm that no matter what he said in his textbook, no matter what von Neumann had proved, an ordinary reality interpretation of quantum theory was possible. In 1952 Bohm constructed such a model for the electron. In Bohm's model the electron is a particle, having at all times a definite position and momentum. In addition, each electron is connected to a new field—the so-called "pilot wave"—which guides its movement according to a new law of motion. Both wave and particle are real—no fictitious proxy waves here—but the pilot wave is invisible, observable only indirectly via its effects on its electron. In Bohm's model, quantumstuff is not a single substance combining both wave and particle aspects but two separate entities: a real wave plus a real particle.

The pilot wave, acting as a sort of probe of the environment, changes its shape instantly whenever a change occurs anywhere in the world. In turn, pilot wave communicates news of this change to electron, which alters its position and momentum. When you make one kind of measurement, the pilot wave has one form; when you make another kind of measurement, this wave takes another form. For different kinds of measurement the electron takes on different attributes, because its pilot wave is different. Thus Bohm's model simulates the electron's contextual behavior using entities (pilot wave plus particle) whose attributes are not contextual. The electron's attributes are innate, but seem to be contextual because the omnipresent pilot wave renders these attributes delicately and immediately responsive to every detail of its environment, including the type of measurement you are preparing to make upon it.

The notion that quantum theory could be explained in terms of an ordinary wave guiding an ordinary particle originated with Prince de Broglie in the late twenties. Encountering severe mathematical difficulties, de Broglie abandoned his model in favor of the reigning Copenhagen doctrine until David Bohm, a quarter century later, showed how the wave-plus-particle concept could be made to work to produce a consistent picture of the quantum facts.

Bohm's pilot wave model revived neorealist hopes that quantum theory might be explained in terms of ordinary objects. However, Bohm's model is plagued with a peculiar affliction. In order for it to work, whenever something changes anywhere the pilot wave has to inform the electron instantly of this change, which necessitates faster-than-light signaling. The fact that superluminal signals are forbidden by Einstein's special theory of relativity counts heavily against Bohm's model, but he was never able to rid it of this distressing feature. Because of its somewhat contrived nature and the presence of superluminal influences, Bohm regarded his model as a mere beginning, as a concrete demonstration that an ordinary reality model of quantum reality was indeed possible. Encouraged by this initial success, Bohm continued to look for a better picture of the reality he was convinced lay behind the quantum facts.

BELL'S INTERCONNECTEDNESS THEOREM

In 1964 Irish physicist John Stewart Bell, working at CERN, the European accelerator center in Geneva, Switzerland, took sabbatical leave from the fast-paced world of high-energy physics to explore the byways of quantum reality. The first question Bell asked was: how was Bohm able to construct an ordinary reality model of the electron when von Neumann had proved that nobody could ever do such a thing? Bohm's model actually did what it claimed: it duplicated the results of quantum theory using a reality made of nothing but ordinary objects. So the fault must lie not in Bohm's model but in von Neumann's proof.

Bell carefully studied this proof in its original version and several variations which other theorists had worked out since the publication of the quantum bible. He was able to find the loophole which permits Bohm's ordinary reality model to exist.

Von Neumann and his colleagues had shown that any scheme in which ordinary objects combined "in reasonable ways" could not reproduce the results of quantum theory. Bell showed that von Neumann's notion of reasonable ways was unnecessarily restrictive. In particular, von Neumann would not have considered "reasonable" electrons which could adjust their attributes via an invisible field that can sense the configuration of the measuring device. Bohm's model, which is based on such context-adaptable electrons, is not "reasonable," hence it evades von Neumann's proof. The fact that thirty years passed before this loophole was discovered is a measure both of the authority of von Neumann and the leisurely pace of quantum reality research.

As he examined von Neumann's proof, Bell wondered whether a truly ironclad argument could be constructed which would set firm limits on the sorts of realities that could underlie the quantum facts. While visiting SLAC—Stanford Linear Accelerator Center—Bell discovered such a proof, which has since become known as Bell's theorem. The unusual demands Bell's theorem makes on reality gives us our clearest picture to date of the irreducible strangeness of the quantum world.

Arguing from quantum theory plus a bit of arithmetic, Bell was able to show that any model of reality whatsoever—whether ordinary or contextual—must be *non-local*. Bell's theorem has since been proved entirely in terms of quantum facts; no reference to quantum theory is necessary. In its most up-to-date version Bell's theorem reads: The quantum facts plus a bit of arithmetic require that reality be non-local. In a local reality, influences cannot travel faster than light. Bell's theorem says that in any reality of this sort, information does not get around fast enough to explain the quantum facts: reality must be non-local.

Suppose reality consists of ordinary objects which possess their attributes innately. Bell's theorem requires for such a world that its objects be connected by non-local influences. Bohm's model is an example of such a world. In this model an invisible field informs the electron of environmental changes with a superluminal response time. Bell's theorem shows that the faster-than-light character of Bohm's pilot wave is no accident. Without faster-than-light connections, an ordinary object model of reality simply cannot explain the facts.

Suppose reality consists of contextual entities which do not possess attributes of their own but acquire them in the act of measurement, a style of reality favored by Bohr and Heisenberg. Bell's theorem requires for such entities that the context which determines their attributes must include regions beyond light-speed range of the actual measurement site. In other words, only contextual realities which are non-local can explain the facts.

Bell's theorem proves that any model of reality, whether ordinary or contextual, must be connected by influences which do not respect the optical speed limit. If Bell's theorem is valid, we live in a superluminal reality. Bell's discovery of the necessary non-locality of deep reality is the most important achievement in reality research since the invention of quantum theory.

Though motivated by quantum theory, Bell's theorem has deeper roots. Von Neumann's proof, for instance, depends on the truth of quantum theory; Bell's theorem does not. As we shall see, to prove Bell's theorem all you need are the facts plus a little arithmetic. The relevant facts are not in question; John Clauser measured them at Berkeley in 1972. Though today's quantum theory shows no sign of weakness, someday it may collapse. Bell's theorem will survive its demise and impose non-locality on quantum theory's successor. Because Bell's theorem makes contact with a general feature of reality itself, it foretells the shape of all future physical theories.

FEYNMAN'S VERSION OF QUANTUM THEORY

In the late forties, while Bohm was writing his popular textbook on the Copenhagen interpretation, Richard Feynman, then a professor at Cornell, discovered a fourth version of quantum theory called the "sum-over-histories" approach. Although it makes the same predictions as the other three quantum theories, the Feynman variation is more than just another language and cannot be reached by a Dirac transformation. It is a fundamentally new way of looking at quantum theory.

Quantum Theory #4: Heisenberg represented it as a matrix, Schrödinger as a wave; Feynman represents quantumstuff as a *sum of possibilities*. Everything that might have happened influences what actually does happen. Feynman's quantum possibilities are different from classical probabilities. Classically the more ways an event can happen, the more probable its occurrence. In quantum theory, possibilities have a wavelike nature that allows them to cancel, so increasing the number of quantum possibilities does not always make an event more probable.

To calculate an electron's fate, Feynman adds up all its possible histories. In the peculiar quantum manner, many histories will cancel. Whatever is left represents what will actually happen—expressed as a pattern of probabilities.

Feynman's sum-over-histories approach is particularly useful for carrying out complex quantum calculations. To predict what will happen in a particular situation, a theorist ranks classes of possible histories in terms of their relative importance and adds the biggest ones first. To keep track of which histories have already been summed, Feynman invented the celebrated Feynman diagrams. Each diagram is shorthand for an entire class of possible histories. So pervasive are the Feynman hieroglyphs that instead of talking about "summing over histories," physicists usually speak of "summing over diagrams."

My exposition of quantum theory in Chapter 6 is inspired largely by Feynman's picturesque sum-over-histories approach. Feynman's way of doing things is original and daring. At a recent conference, Feynman's colleague Freeman Dyson recalled his first impressions of this unusual approach to quantum theory: "Thirty-one years ago, Dick Feynman told me about his 'sum over histories' version of quantum mechanics. 'The electron does anything it likes,' he said. 'It just goes in any direction, at any speed, forward or backward in time, however it likes, and then you add up the amplitudes and it gives you the wave function.' I said to him 'You're crazy.' But he isn't."

This account of the highlights of quantum reality research is necessarily brief and incomplete. For the reader in search of more detail, I recommend Max Jammer's excellent book *The Philosophy of Quantum Mechanics.*

Physicists did not willingly give up ordinary reality to wallow in dozens of bizarre and contradictory pictures of the world: experiments pushed them into the quantum soup. We turn now to those persuasive experiments, physicists' court of last resort, the troublesome quantum facts.

4

Facing the Quantum Facts

I remember discussions with Bohr which went through many hours till very late at night and ended almost in despair, and when at the end of the discussion I went alone for a walk in the neighboring park I repeated to myself again and again the question: "Can nature possibly be as absurd as it seemed to us in these atomic experiments?"

Werner Heisenberg

Physicists, for all their odd notions, are basically a conservative lot. They would have been content, most of them, to dwell in the solid classical world created by the great scientists of the Victorian Era and leave wild speculations concerning the nature of things to the science fiction chronicles of Jules Verne. However, new quantum facts forced physicists to admit that the world almost certainly rests on some bizarre deep reality. If scientists routinely contact facts which reveal such outlandish realities, life in a modern physics lab must be pretty unusual.

One imagines Max, the famous quantum physicist, deciding on Monday morning to face the quantum facts. Donning quantum-resistant body armor, he climbs inside his bubble chamber, waves goodbye to the workaday world, and prepares to enter the mysterious realm of the quantum. Alone in the dark, Max checks his life-support system and the crucial flyback circuit that returns him to ordinary reality. Then, taking a deep breath, he pulls the switch.

Max suddenly drops through the world's phenomenal surface into deep quantum reality. Holy Heisenberg! Centuries of Newtonian certainties vanish in an instant. Solid objects melt into the undivided wholeness as he enters the Place Without Separation. Max mixes with the mystery when his subject/object membrane dissolves. In tune with totality, Max creates a new universe faster than light wherever he turns his omnipotent gaze.

What's it like down there? Max's sister Maxine says it feels just like Schrödinger's equation, only more so. You've got to see it to believe it. Behind the high-security fences of Max's quantum lab, consciousness creates reality, quantum logic is spoken exclusively, and for the trip home you have your choice of a billion different universes.

CINDERELLA EFFECT: THE ORDINARINESS OF QUANTUM FACTS

Sad to say, physics labs are not so exciting. Despite the outlandish realities invented to explain them, quantum facts consist of quite ordinary events; these quantum experiments are remarkably commonplace, especially when compared with the extravagant claims of the quantum realists. Even our clearest factual window into deep reality—the celebrated EPR experiment which validates Bell's interconnectedness theorem—is, as we shall see, absolutely ordinary.

All quantum experiments consist of commonplace events, a fact I call the Cinderella effect. The world may really be as strange as some physicists say, but it does not flaunt this strangeness, evidently preferring to hide its magic—like Cinderella—in humble guise. The Cinderella effect itself is a subtle example of quantum weirdness: why does nature employ such extraordinary realities to keep up merely ordinary appearances?

Niels Bohr in his Copenhagen interpretation of quantum theory gave prominent place to the Cinderella effect when he insisted that all quantum experiments be described in ordinary—Bohr called it "classical"—language:

"However far the phenomena transcend the scope of classical physical explanation, the account of all evidence must be expressed in classical terms . . . The account of the experimental arrangement and of the results of observation must be expressed in unambiguous language with suitable application of the terminology of classical physics."

In other words, although the explanation of quantum facts is far from ordinary, the facts themselves are made from the same kinds of events as prequantum facts—unit acts as unremarkable as those of everyday life. Bohr was one of the few quantum theorists to emphasize the ordinariness of quantum fact. Bohr believed that ordinariness is built into human modes of perception so that all future quantum facts would likewise be ordinary. Humans are fated to experience the quantum world secondhand: we will never, like Max, enjoy direct experience of quantum reality.

Sixty years of experiments agree with Bohr. Today's state of the quantum art is such that we cannot directly experience quantum reality. All human experiences—or at least all physics experiments—are ordinary, not quantum, in appearance. Whether our reliance on classical modes of perception is a permanent feature of the human condition remains to be seen. We humans are ingenious animals, perhaps too ingenious for our own good. Since "reality has consequences," we might anticipate that if one of these quantum realities is "really real," we will eventually figure out how to experience it directly: Max's bizarre quantum lab may not be so farfetched in the future. Now, we see quantum reality through a glass darkly, but then, face to face. However, since all of today's quantum facts are admittedly ordinary, what basis do physicists have for their outrageous claims regarding deep reality?

Quantum facts are indeed ordinary. But quantum theory—the only complete explanation we possess of these facts—is decidedly non-ordinary. Since quantum theory fits the facts exactly, many physicists are sure that it bears some relationship to reality itself: such a perfect match between theory and fact is no accident. Physicists come up with different pictures of quantum reality depending on what aspects of quantum theory they decide to take seriously and which parts they discard as mere mathematical figures of speech.

Quantum reality doesn't show up *directly* in the quantum facts: it comes indirectly out of the quantum theory, which perfectly mirrors these facts. Before examining this theory which supports so many odd realities, let's take a look at some of the facts it so successfully explains.

The simplest conceivable quantum experiment consists of a source of quantumstuff, a quantumstuff detector, plus something to put in between that alters quantumstuff in a systematic way.

TESTING A QUANTUM ENTITY

As a typical quantum entity I choose the electron, the first "elementary particle" to be discovered—by Englishman Joseph J. Thomson in 1897. Despite modern attempts to split it into finer bits, using energies a hundred billion times greater than those that hold the atom together, the electron remains steadfastly elementary. An electron, so it seems, simply doesn't have any parts.

Inside a TV set, astronomical numbers of electrons detect, decode, and display information carried into your home via electromagnetic waves vibrating ten million times slower than waves of visible light. The heart of a TV set is its picture tube.

Inside the picture tube, a beam of electrons paints a new TV image on the phosphor screen sixty times a second. For our quantumstuff source and detector, we will borrow from a picture tube its electron gun and phosphor screen.

The electron gun contains a metal filament heated to boil off electrons, and some charged metal cylinders which accelerate them to high velocities. A TV electron gun—about six inches long—is a miniature linear accelerator. Inside a color TV, electrons strike the screen with an energy of about 25 thousand volts. Electrons in Stanford's two-mile-long linear accelerator attain energies a million times greater.

Knobs behind your TV set control the action of the electron gun. When we remove the electron gun for our quantum experiment, we will also take two of these knobs: the brightness control, which varies the intensity—electrons per second—of the beam; and the high voltage control, which varies the electron's momentum—how fast it's moving when it strikes the phosphor screen.

The active ingredient of the phosphor screen is a solid substance—the phosphor—ground to the consistency of face powder, mixed with glue, and spread thin against the back of the glass faceplate. As its name implies, a phosphor is a molecule that gives off light. A phosphor molecule usually resides in its "ground state"—its state of lowest energy, symbolized by P. When a phosphor gains energy—by being hit by a fast electron, for instance—it transforms into an "excited state," symbolized by P* (pronounced "P star"):

$$P + Energy \rightarrow P^*$$

The phosphor doesn't stay in this excited state for long. It wants to get rid of its excess energy and return to its ground state. For a phosphor, the easiest way to lose energy is to emit a photon of light:

$$P^* \rightarrow P + light$$

Now the phosphor is back in its ground state, ready to be excited again, and some of the electron's energy has been converted to light. The overall action of a phosphor looks like this:

$$P + Energy \rightarrow P + light$$

Just as a chemical catalyst facilitates chemical reactions without being itself changed or used up, so a phosphor is a kind of energy catalyst remaining unchanged as it turns some of an electron's kinetic energy into light.

Phosphors are particularly useful as quantum detectors because they are sensitive enough to respond to a single quantum (in this case an electron) and are not particular about where their excitation energy comes from. A phosphor glows when struck by electron, but proton, pion, or any other charged particle will do as well. A phosphor can even be excited by light—that is, by a photon. Turning light into light may not seem very practical, but makes sense if the input light is invisible—infrared or X rays, for instance. Phosphors that turn invisible radiation into visible light make good X-ray viewing screens.

A phosphor can be excited by as little as one volt of energy. A quantum particle carrying several thousand volts can excite many phosphor molecules, which appear as a bright flash easily visible to the naked eye. Phosphors are sensitive, fully reusable quantum detectors which signal the presence of a quantum particle with a flash of light. So that we can discuss the quantum measurement process in an orderly fashion, all my detectors use phosphor molecules as their primary sensors.

THE ELECTRON'S PARTICLE NATURE

For our first quantum experiment, we aim the electron gun at the phosphor screen and observe a small dot of light where the beam strikes the screen. The intensity of this dot goes up as we increase the number of electrons by turning up the brightness or as we increase the electron energy by turning up high voltage. As we manipulate these controls, the spot size remains the same.

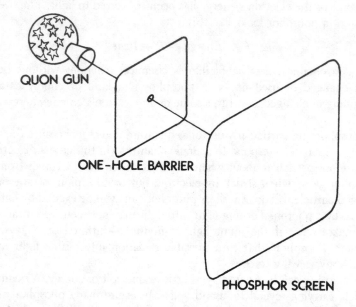

QUON GUN

ONE-HOLE BARRIER

PHOSPHOR SCREEN

FIG. 4.1   *Electron diffraction experiment. An adjustable source (quon gun) shoots electrons at a phosphor screen through a tiny circular hole. A wave would spread as it went through the hole; a particle cannot spread but must take one particular path. Will the electron choose the wave or the particle option?*

Turning down brightness decreases the spot's intensity. At low-beam intensity, we can see the effect of the individual electrons which make up the beam: the spot does not shine constantly now but sparkles as each new electron excites a clump of phosphors. At very low brightness we can watch the arrival of single electrons as they strike the TV screen. This experiment gives tangible evidence for the electron's particle nature: electrons are particles because you can count them.

How small are these little particles of electricity? Physicists have attempted to determine the electron's size by using another electron as a probe. Two electron beams are accelerated, then bent with magnets so they meet head on. Electrons don't actually have to touch in order to interact, because each one is surrounded by its own electric field. At large separations electrons scatter via these fields. Physicists hope to measure their intrinsic size by pushing them closer and closer together (by increasing the accelerator energy) until deviations from pure-field scattering occur, indicating that the electrons themselves are beginning to touch.

Present-day accelerators can probe distances as small as 10^{16} cm—a thousand times smaller than a proton's diameter. At these tiny separations, electrons still show nothing but pure-field scattering. If the electron has any size at all, it is smaller than we can measure. Some physicists conjecture that the electron is a *point particle* whose intrinsic size is zero! All truly elementary particles, they imagine, are likewise mathematical points. Only composite entities such as atoms, built up of elementary particles, will show a structure or have a definite size. None of the current candidates for elementarity—i.e., quarks, leptons, or gluons—shows any detectable structure, which agrees so far with this point-particle conjecture.

Electron is an example of quantumstuff which possesses both wave and particle properties. The diameter of the electron-as-particle is very small: it may be zero. We can actually "see" these electron "particles" arriving one at a time on the TV screen; evidence for the electron's particle nature is clear and indisputable. Next we look at the evidence for its wave nature.

THE ELECTRON'S WAVE NATURE

As our "wave probe" we insert a small circular iris between the electron gun and the luminescent screen, positioning it so that the electron beam goes through the hole. We make the hole smaller, squeezing down on the electron beam. At first the electron beam gets smaller as the hole shrinks, as shown by the smaller spot on the screen. However, beyond a certain point the spot refuses to shrink and actually begins to get larger. As we continue to decrease the iris size, the image on the phosphor screen expands until it is no longer a spot but a series of bright and dark rings resembling an archery target.

When we contract the hole further, this target figure enlarges, its bulls-eye expanding to fill the entire screen, till finally—when the iris is very small

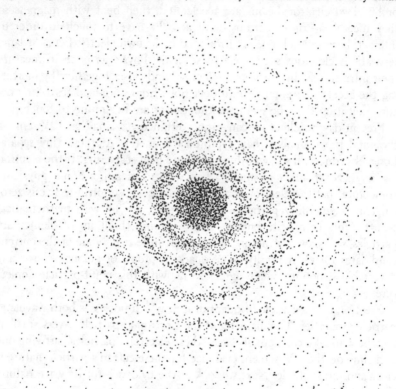

FIG. 4.2 *The outcome of electron diffraction experiment: each electron makes a tiny flash of light; collectively these flashes form a spread-out design—the so-called Airy pattern. In this experiment the electron reveals both its wave and particle aspects.*

—the screen is covered with a uniform glow. As the hole continues to contract, the phosphor screen remains uniformly illuminated but the light intensity decreases. Finally, when the hole closes completely, the screen goes dark.

We open the iris again, adjusting it so the electrons reform the target-shaped pattern. Physicists are familiar with this particular pattern of concentric rings; it was first explained in 1835 by George Biddell Airy, the British Astronomer Royal. Airy calculated mathematically the pattern a wave would make if constrained to go through a circular hole. His calculation applies to any kind of wave—sound wave, water wave, light wave, or quantumstuff wave. Airy's pattern is for waves a telltale fingerprint. When you put something through a circular hole and it makes an Airy pattern, then you know you're dealing with a wave.

The Airy pattern's overall shape is caused by wave diffraction: the ability of a wave to bend around corners. The light and dark rings are caused by wave interference: the recurrent alternation of intensity as wavelets traveling different paths arrive in or out of phase. Airy showed that the total pattern can be explained as the interplay of diffraction and interference effects.

Airy's calculation describes all the details of the target figure—the diameter of each dark ring and the intensity of each light circle. In particular the angular diameter θ (in degrees) of the central bulls-eye (Airy's disc) is given, in terms of the wave's wavelength L and the iris diameter d by the simple formula:

$$\theta = 70(L/d)$$

Notice that this formula describes an inverse relation between hole size d and the angular size of the bull's-eye θ agreeing with the experimental observation that the pattern gets bigger as the hole shrinks. We also see that when the hole is the same size as the wavelength, Airy's disc subtends an angle of 70 degrees, a fact which allows us to calculate the wavelength of the electron from the size of the Airy pattern.

Airy didn't have electrons in mind when he did his calculation—they were yet to be discovered. He was trying to explain the pattern that light makes when it goes through a small hole. You can see the optical Airy pattern for yourself by punching a tiny hole (0.1 mm) in a piece of aluminum foil, holding it up to your eye and viewing a point source of light (the sun's too big) such as a distant streetlamp.

In the nineteenth century, Airy's explanation of this pattern as a signature of wave action provided a key piece of evidence for the wave nature of light. One hundred years later, physicists regard the electron's Airy pattern as twentieth-century evidence for the wave nature of matter.

An electron seems to possess contradictory attributes. As a particle, it must be localized in space, cannot be split apart, and retains its identity in collisions with other particles. As a *wave*, it spreads over vast regions of space, is divisible in an infinity of ways, and merges completely with other waves it happens to meet.

A purely particle theory cannot explain the Airy pattern; a purely wave theory cannot explain the flashes on the screen. The electron is in reality neither particle nor wave, but an entity entirely new to human experience which exhibits the properties of both. The electron is pure quantumstuff.

We've carried out this experiment with electrons, but every other quantum entity behaves the same way. Photons, quarks and other elementary particles will likewise show an Airy pattern built of little flashes if you put them through a tiny hole. To see their quantum nature, merely replace the electron gun with one that shoots photons or quarks.

Most physicists believe that ordinary objects—baseballs, tomatoes, Mack trucks—would in principle exhibit quantum wave properties under the right conditions, but their extremely short de Broglie wavelengths make such effects impossible to observe in practice.

Everything in the world is pure quantumstuff, a physical union of particle and wave. The particle aspect of light waves is called "photon"; the particle aspect of gravity is called "graviton"; the particle aspect of the strong nuclear force is called "gluon." No term exists for a generic quantum object. I propose the word "quon." A quon is any entity, no matter how immense, that exhibits both wave and particle aspects in the peculiar quantum manner.

What's an experiment without knobs to wiggle? Let's see what happens to the Airy pattern as we change the electron gun's high voltage and brightness controls.

First the high voltage control. As we turn up the voltage, the Airy pattern gets smaller, eventually shrinking back down to a tiny spot. Turning down the voltage makes the pattern expand. Evidently the high voltage control affects the electron's wavelength: the higher the voltage, the shorter the wavelength. In a TV tube, high voltage corresponds to high electron momentum. We can use the observed variation in Airy pattern size with voltage to discover the relationship between the electron's momentum and its wavelength. For each high voltage setting, we calculate the electron's wavelength L from Airy's formula and the observed size of the Airy disc. The electron's momentum p is related to its voltage by a well-known classical formula. The experimental relation obtained in this TV-based diffraction experiment between an electron's momentum and its wavelength is particularly simple:

$$p = h/L$$

where h is a constant which appears in almost every quantum relation—Max Planck's ubiquitous constant of action.

Expressed in terms of spatial frequency k rather than wavelength L, this relationship becomes:

$$p = hk$$

Whether to express the electron's wave aspect in terms of wavelength or spatial frequency is a matter of convenience. What's important is that these electron diffraction experiments demonstrate an extremely simple connection (involving Planck's constant) between the particle property momentum and a property (wavelength or spatial frequency) heretofore associated only with waves.

This notion that the electron, then considered a mere particle, also possessed wave aspects was first proposed by French physicist Louis de Broglie in 1924. De Broglie predicted the above momentum/wavelength relation, which was subsequently verified by Americans Davisson and Germer in electron diffraction experiments similar in principle to our one-hole TV display.

ARE ELECTRON WAVES "CROWD WAVES"?

The fact that the electron shows both wave and particle properties is not in itself peculiar. What's strange is how these properties coexist. We don't find it remarkable that water waves consist of particles—molecules of water—that collectively behave like a wave. Sound is another example of a wave riding on a collection of particles. I call such waves made of particles "crowd waves." To make a crowd wave, many molecules must be packed together. A bell ringing in a sealed jar makes no noise when there's not enough air inside to support sound waves.

The crucial test for crowd waves is dilution. We can make a crowd wave disappear by decreasing the number of molecules in the crowd. Pumping the air out of the jar silences the bell.

We dilute our electron beam by turning down the brightness. First adjust all controls for a big clear Airy pattern. Then reduce the brightness. Will the Airy pattern go away once the number of electrons gets small, revealing the electron wave to be a mere "crowd wave"? Don't bet on it.

As we turn down the brightness, the number of electrons striking the screen decreases. Glow turns to sparkle, breaking up gradually into individual flashes. Finally only about one electron per minute hits the TV screen. At this low brightness, most of the time the TV tube is empty. Certainly this drastic electron deficit has eliminated any possibility of crowd waves. But it's also eliminated the Airy pattern. All we see on the screen is a single flash every minute or so.

To record possible long-term patterns in these isolated flashes, we press a piece of photographic film against the glass. A few weeks later we return and develop the film. The first physicist to perform this kind of quon-dilution experiment, Englishman G. I. Taylor, went sailing on the Thames while his flashes accumulated. Returning to the lab, we develop the film. We see thousands of dots, each a token of a single electron. But rather than being scattered at random, they form exactly the same Airy pattern as the high-intensity beam. Electrons evidently behave like waves no matter how much you dilute them. Electrons are definitely not crowd waves like sound and surf.

This quon dilution experiment shows that, although it appears on the screen as a particle, each electron by itself travels from gun to screen as though it were a wave. The electron manages its contradictory aspects by assuming them one at a time. Whenever it's being observed, an electron always looks like a particle—and a mighty small one at that. In between observations, the same electron spreads out like a wave over large regions of space. This alternation of identities is typical of all quantum entities and is the major cause of the reality crisis in physics.

WAVE/PARTICLE COEXISTENCE

The world is made entirely of quons that behave like this electron. How shall we explain such an entity? It acts like a particle whenever we look. In between looks it acts like a wave. Because measured electron is radically different from unmeasured electron, it appears that we cannot describe this quon (or any other) as it is without referring to the act of observation.

If we ignore observations for the moment, we might be tempted to say that an electron is all wave, since this is how it behaves when it's not looked at. However this description ignores the massive fact that every observation shows nothing but little particles—only their patterns are wavelike. If we say, on the other hand, that between measurements the electron is really a particle, we can't explain the quantum facts. How does each electron on its own know how to find its place in the (wavelike) Airy pattern? What does a single electron "interfere with" to produce Airy's dark rings?

The essential difficulty in describing quantum reality is that unmeasured quons seem to behave in a totally different manner from measured quons, and that neither behavior by itself is enough to explain how the world itself behaves. Quantum realists would like to be able to give a single description of the world as it is, independent of how it seems when we look at it. Of course such a description should explain how the world appears when measured, but the measurement act should be a subordinate part of any model of reality, not an essential feature.

However, the quantum facts give us not one description but two—each one separately inadequate, and both together contradictory. Moreover the knot that connects these two descriptions is the act of observation; leave out observation and neither description makes sense.

Physicists customarily appeal to experiment to settle matters of principle. Using a TV tube, we looked at some of the electron's peculiar quantum properties. Certainly there must be other experiments we could do that would tell us more than we could learn from this single setup. Perhaps by making other kinds of measurements we could learn what the electron's actually doing in the TV tube when it's "unobserved."

THE UNCERTAINTY PRINCIPLE PROTECTS WAVE/PARTICLE COEXISTENCE

Whenever we perform other experiments we indeed get new information, but it's never enough to solve this problem. All experiments show the same kind of unmeasured wave, measured particle duality. An instrumental barrier seems to exist which prevents probing the quantum world deep enough to resolve the wave/particle question in favor of one or the other modes of being. Too close scrutiny of a quon's behavior is blocked by the distinctively quantum feature of conjugate attributes.

In classical physics, all of an entity's attributes are in principle accessible to measurement, with a precision limited only by the experimenter's ingenuity. A quon's measurement situation is quite different. You can measure a single quantum attribute as accurately as you please, but such a measurement inevitably produces imprecision in some other quantum attribute.

Unlike the independent attributes of classical physics, quantum attributes seem joined to other attributes, at least during the act of measurement. A systematic study of which attributes are linked during measurement shows that quantum attributes always come in pairs. Each attribute A enjoys a special relationship with another attribute V called its "conjugate attribute." The fact that A and V are conjugates limits how accurately you can know them both. If you measure A with good precision, then you must accept a poor precision when you go to measure attribute V. In general the mutual precisions of attributes A and V are constrained by a relationship of the form:

$$\Delta A \, \Delta V > K$$

where ΔA and ΔV are variances in the experimental values of conjugate attributes A and V. This relation requires that no matter how skillful the experimenter, the product of the precisions ΔA and ΔV of his measurements can never be less than a certain constant K. Of course if the experiment is not carried out with perfect care, the variances may exceed this limit but they can never be less. For a perfect quantum measurement, the product of two conjugate variances equals K.

This measurement restriction holds for all dynamic attributes—any attribute that's not constant. In the extreme case of measuring one attribute with perfect precision $(\Delta A = 0)$, all knowledge of its conjugate attribute is forfeited $(\Delta V = \infty)$. Because of this fundamental measurement restriction on conjugate attributes, there is a sense in which *half of a quon's attributes are always hidden from view*. This restriction on the mutual measurement precision of certain attributes is just sufficient to prevent you from devising an experiment that would show you what's really there and decisively resolve the wave/particle dilemma.

The most important conjugate attributes (and the first to be discovered) are a quon's position and momentum. The paired relationship which restricts their mutual measurement precision is known as Heisenberg's uncertainty principle:

$$\Delta p \, \Delta x > h$$

where Δp is the uncertainty in momentum, Δx the uncertainty in position, and h is Planck's constant. Heisenberg's principle says that if you measure position accurately you must sacrifice an accurate knowledge of momentum. A relationship of the Heisenberg kind holds for all dynamic attributes, and prevents anyone from resolving the quantum reality question via the clear light of experiment. The Heisenberg relations guarantee that any experiment will contain a blind spot just big enough to hide the solution to the wave/particle riddle.

As well as demonstrating the electron's wave/particle nature, the Airy experiment illustrates Heinsenberg's principle. This experiment can be construed as an attempt to measure the sidewise position of the electron with a precision equal to the diameter of the hole in the probe plate. Immediately after the electron goes through the hole, we know its position to within a precision $\Delta x = d$.

Airy's formula shows that the smaller the hole, the more the beam diffracts as it goes through. This means that, for a small hole, the beam acquires a large spread in sidewise momentum because of this position measurement. The spread in momentum induced by a hole of a particular size can be calculated by Airy's formula. Putting all this information together gives an uncertainty relation precisely equal to the Heisenberg limit:

$$\Delta p \, \Delta x = h$$

For the Airy experiment, the mutual spreads in precision of these two conjugate attributes is equal to Planck's constant. The Airy experiment is an example of a perfect quantum measurement. It represents the most anyone can find out about both the sidewise position and the sidewise momentum of an electron beam.

The flavor of the strange quantum world permeates these simple experiments. It's a world that's wavelike when unobserved, particlelike upon observation: a world whose attributes come in pairs which jointly resist close examination. Quantum theory accounts for these facts and many more. However, rather than resolving the quantum reality question, quantum theory merely deepens it.

Quantum theory, because it precisely mirrors the quantum facts, possesses the same qualities that prevent us from building a consistent observer-free picture of reality from the quantum facts. In addition quantum theory brings puzzling features of its own (wave function collapse and phase entanglement, for instance) whose relation to what really goes on in the world is highly dubious.

Running parallel to the quantum facts, quantum theory represents unmeasured quons as waves and measured quons as particles. Furthermore it regards these unmeasured waves not as real waves but merely as waves of probability. On account of its indirect method of representation, this theory, highly successful in practice, seems farther removed from reality than the experiments it so accurately predicts. However, barring a sudden breakthrough into Max's direct way of contracting quantum reality, quantum theory is the best clue we possess concerning the real nature of the world we live in.

So we can better appreciate the probability waves with which quantum theory characterizes the world in its unmeasured state, I review in the next chapter some familiar properties of ordinary wave motion.

5

Wave Motion:
The Sound of Music

The clocks on the tower of the Ferry building said that it was 5:15
—they were running a little fast. But it was to be months before
those hands moved any farther, for at that instant the earthquake
struck . . . Jesse Cook, who was later to become police commis-
sioner, remembered hearing a deep rumbling in the distance, "deep
and terrible" in his words. And then looking up Washington
Street, he actually *saw* the earthquake coming. "The whole street
was undulating. It was as if the waves of the ocean were coming
toward me, and billowing as they came."

> William Bronson, reporting on the
> San Francisco earthquake

Using ordinary waves in unusual ways is the secret of quantum theory.

All waves, no matter how exotic, are built on a common plan and take
their orders from the same rulebook. Although physicists connect quantum waves with facts in an innovative way, the quantum waves themselves
follow the same old-fashioned rules as waves in your bathtub. In this

chapter I look at fundamental behavior common to all waves. Later we'll examine this same behavior in waves of quantumstuff.

Waves take their character from what's doing the waving: water and air are waving in the case of surf and sound. According to James Clerk Maxwell, light is a vibration of electric and magnetic fields. Quantum waves, as we shall see, are oscillations of possibility.

Since a wave vibrates both in time and space, to follow it we must keep track of two kinds of motion. One way to do this is to make two separate pictures—one in which we stand still in space watching the wave change in time; the other in which time stands still and we look at how the wave changes in space. We freeze time to get a wave's spatial picture; we freeze space to get its temporal picture.

A wave's fundamental scale is its amplitude, which measures the deviation of its physical variable from the rest state. Another important wave measure is intensity, which is proportional to amplitude squared. For all waves except quantum waves, intensity measures the amount of energy a wave carries at every point. Quantum waves carry no energy at all; for this reason they are sometimes called "empty waves." A quantum wave's intensity (amplitude squared) is a measure of probability.

A wave takes any form it pleases: some waveforms are one of a kind, others are oscillatory—a parade of identical shapes like a modern production line. Oscillatory waves go through cycles in time and space; their essence is repetition. The time an oscillatory wave takes to go through one cycle is called its "period." Cycle time can also be expressed in terms of frequency: the number of cycles completed in a certain time. Period P and frequency f are equivalent names for the rapidity of a wave's pulsations in the temporal picture.

The space an oscillatory wave spans as it carries out one cycle is called its wavelength. Cycle length can also be expressed in terms of spatial frequency: the number of cycles filling up a certain distance. Wavelength L and spatial frequency k are equivalent names for the repetition rate of a wave's undulations in the spatial picture.

Phase is another important measure of an oscillatory wave. Each point on a wave possess a definite phase which tells how far that point has progressed through the wave's basic cycle. The term "phases of the moon" expresses this same meaning of "phase" as part of a cycle. The phase of a wave governs what happens when two waves meet. Wherever waves of the same frequency (spatial or temporal) come together with identical phases, they are said to be "in phase"; waves whose phases differ by half a cycle are "out of phase."

SUPERPOSITION PRINCIPLE

The meeting of two waves to make a new wave may look complex but what actually goes on is remarkably simple: the new wave's amplitude at every point is just the sum of the amplitudes of each separate wave. *When waves meet, their amplitudes add.* The fact that waves everywhere form such uncomplicated unions is called the "superposition principle." This principle works not just for oscillatory waves but for all waveforms whatsoever.

Ordinary waves obey the superposition principle for small amplitudes, but not when amplitudes get big. Failure of the superposition principle is called non-linearity, and shows up as distortion in hi-fi systems and as turbulence in water waves. A remarkable feature of quantum waves is that they seem to obey the superposition principle without restriction: no matter how complex the circumstances, the amplitudes of quantum waves add, and nothing more. When you get down to the quantum level, wave behavior is simpler than waves in your bathtub.

The superposition principle guarantees that when waves come together, nothing is added or taken away. In particular, when a wave exits such a relationship it takes with it precisely the amplitude it had when it went in. Two waves can cross paths, form a momentary superposition, then continue on their ways entirely unchanged by their encounter—an option not generally available to other forms of being.

Consider a sunny window. Light from many different directions crosses as it goes through the glass. If the passage of light waves through one another changed them in any way, the information they carry would be distorted. Yet, because light waves interact in a "reasonable" way, the scene outside never blurs no matter how bright the light. The superposition principle applied to light waves keeps your windows clear.

Because quantum theory in a certain sense regards the world as made out of waves rather than out of things, quantum entities and their attributes combine according to the rules of *wave addition* rather than the rules of ordinary arithmetic. The superposition principle, which governs how waves add, is as important for the quantum world as arithmetic is for everyday life.

WAVE INTERFERENCE

The superposition principle applied to oscillatory waves requires that when such waves add, the amplitude of the resultant wave depends crucially on *phase relations*.

When two waves add *in phase*, peaks line up with peaks, valleys with valleys to make the resultant wave bigger than its components. If both waves have the same amplitude (1 unit), the combined wave will have twice this amplitude. The way that amplitudes add in phase might be symbolized by:

$$1 \oplus 1 = 2 \text{ (amplitudes in phase)}$$

where $\oplus$ is a symbol I've invented to stand for *wave addition*.

When two waves add *out of phase*, peaks line up with valleys to decrease the amplitude of the resultant wave. If each input wave had the same amplitude (1 unit), the combined waves *exactly cancel*, resulting in a wave with zero amplitude. The way in which wave amplitudes add out of phase might be symbolized thus:

$$1 \oplus 1 = 0 \text{ (amplitudes out of phase)}$$

When the phase lies somewhere in between these two extremes, the combined amplitude likewise falls in the middle. Wave addition for *arbitrary phase* might be symbolized like this:

$$1 \oplus 1 = 0 \text{ to } 2 \text{ (amplitudes for arbitrary phase)}$$

When two waves with equal amplitude come together, the amplitude of the combined wave can be anywhere between zero and twice the amplitude of a single wave. The critical factor which decides the outcome of this peculiar wave arithmetic is the waves' relative phase.

Phase is a matter of timing: how soon will the next wave peak get here? These simple examples show the importance of the phase variable for wave addition. If two waves arrive on time, the surf is two feet high; if one wave is half a cycle late, the ocean is mysteriously calm.

This ability of two waves to augment or diminish each other depending on their phase difference is called interference: a particularly unfortunate choice of words since the superposition principle assures us that the one thing waves do not do is "interfere." Like customers making bank deposits, waves add or subtract their amplitudes with complete indifference to another wave's presence. A word which does not suggest hindrance, such as "concurrence," might have been a better choice but several hundred years of tradition have sanctified "interference" as the official name for the phase-sensitive union of oscillatory waves.

The extreme case of waves meeting precisely in phase to achieve maximum enhancement is called "constructive interference." Out-of-phase superposition is called "destructive interference."

Destructive interference finds practical use in anti-reflective coatings of camera lenses. Some of the light striking a lens always reflects back. Reflections at every glass-air interface in today's multielement lenses would result in serious light loss and many extraneous images. These reflections are reduced by coating each glass surface with a transparent film just a quarter wavelength thick. Now light reflects at two interfaces—where the coating meets the air and where it meets the glass.

It may seem that doubling the number of reflective surfaces will only make matters worse, but destructive interference comes to the rescue. Each new surface, because of its critical spacing, produces a wave which is out of phase with the original reflection. In practice, complete destructive interference can be achieved for only one color. The other colors, diminished but not destroyed, give coated lenses their distinctive purple sheen.

WAVE ENERGY

The largest recorded tidal wave—more than 200 feet high—appeared at Valdez, Alaska, in 1964. Four-foot breakers are not uncommon at the beach. Was the great wave of Valdez only fifty times more powerful than everyday surf?

A wave's amplitude measures how big it is, but grossly underestimates the wave's destructive power. A wave's external effect depends on the energy it carries, which is proportional to the wave's intensity (amplitude squared). *Wave energy goes as amplitude squared.* When you double a wave's amplitude, you quadruple its energy content.

Although a quantum wave possesses no energy, its intensity (amplitude squared) does not lack a physical interpretation. For any quantum wave, amplitude squared means *probability.* All that we learn here about the *energy* carried by an ordinary wave is directly applicable to the *probability* carried by a quantum wave. A common feature of energy and probability is that both are conserved. Left to itself, the total energy contained in an ordinary wave never changes; likewise the total probability contained in an isolated quantum wave is constant.

The law of the conservation of energy is a familiar concept: energy can neither be created nor destroyed—only energy transformation is possible. Probability conservation is less familiar: it means, in the Airy experiment, that if one electron goes through the hole, only one electron will hit the screen. Even in the quantum world you never get out any more (or any less) than you put in.

Let's imagine for the moment that we're dealing with ordinary waves for which amplitude squared means energy, and examine what happens to this energy during the process of interference.

A wave of unit amplitude possesses 1^2 or one unit of energy. It carries this one energy unit wherever it goes.

Suppose this wave interacts with another unit amplitude wave and that at a certain location both waves meet in phase. For in-phase wave addition, amplitudes combine like $1 \oplus 1 = 2$. The combined wave possesses 2^2 or four energy units. However, each wave comes in with only one energy unit; together they bring in two units.

Because of the way that waves add, *four* energy units show up at the interference site—two energy units have appeared out of nowhere! The process of in-phase (constructive) interference leads to a local energy surplus: more energy comes out than goes in.

Let's look at the energy balance for destructive interference. Again, each wave brings in one energy unit for a total of two units. For out-of-phase wave addition, amplitudes combine like $1 \oplus 1 = 0$. So the combined wave possesses 0^2, or zero energy units.

Two units of energy go into the interference site; zero units come out—two energy units have vanished into thin air! The process of out-of-phase (destructive) wave addition leads to a local energy deficit: more energy goes in than comes out.

RANDOM PHASE WAVE ADDITION

In addition to adding waves in and out of phase, we could imagine adding them with no regard to phase. A random phase results when a wave's timing fluctuates during the course of the measurement. Since random phase addition involves a disorderly mixture of all possible phases, we might expect the results of wave addition to lie somewhere in between the extremes of perfect *c* and *d* interference. That is, for the random superposition of two unit amplitude waves, we guess that the new wave's amplitude will lie between the extremes zero and two. When we actually carry out a random sum we get:

$$1 \oplus 1 = \sqrt{2} \text{ (amplitude addition for random phase)}$$

Two unit amplitude waves added together with random phases give a combined wave whose amplitude is the square root of two (about 1.4).

As before, let's investigate how the energy behaves during wave superposition. Each unit wave brings in one unit of energy—a total of two units for both. The energy output is just the combined amplitude squared, which is also two energy units. Two units in; two units out. *For random superposition of waves, energy is exactly conserved.*

When two waves meet they form a zebra-striped interference pattern consisting of alternating regions of constructive and destructive interference. In regions of *c*-interference, *more energy* appears than each wave brings in; in regions of *d*-interference, *less energy* appears than each wave brings in. In an interference pattern, local energy is not conserved: there's too much energy in the *c* regions and too little in the *d* regions. However, if we check the accounts carefully we find that no energy is gained or lost overall: energy missing from the *d* regions exactly matches the energy excesses in the *c* regions. Although it redistributes wave energy in an inequitable way, wave interference, like every other physical process, obeys the law of (total) energy conservation.

The Airy experiment shows a typical interference pattern: the Airy pattern's central peak and bright circles are regions of excess energy; its dark rings are regions of energy deficit. More energy flows from the hole into the dark rings than ever reaches the screen: wave energy heading in that direction cancels out by destructive interference. Fig. 5.1 shows the Airy pattern as an alteration of in-phase and out-of-phase wave addition. To see local energy conservation in action, we arrange to randomize the phases of the electrons in the Airy experiment. Instead of the zebra-stripe alternations of energy surpluses and deficits, energy is everywhere conserved. The peaks of the Airy pattern are laid low; its valleys are exalted. The interference pattern changes into a featureless blur. Note, however, that although interference is destroyed by phase randomization, diffraction is not: the pattern spreads out to the same extent as before. The ability of a wave to bend around corners does not depend on its phase.

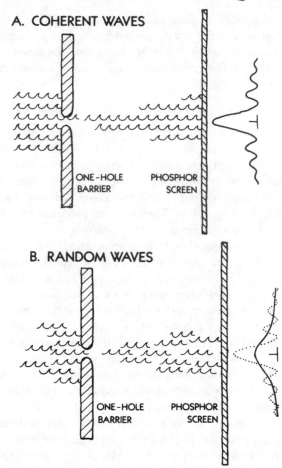

FIG. 5.1 *Coherent and random superposition. A. Orderly (coherent) waves produce an interference pattern. B. Interference disappears when their phases are randomized. Energy is not conserved locally in case A: energy surpluses appear at regions of constructive interference; energy deficits at destructive interference regions. In case B, local energy is conserved everywhere. In both cases total energy (the sum of all local energies) is conserved.*

When ordinary waves superpose with definite phases, energies do not add everywhere. When these same waves superpose with random phases, energies add everywhere.

When quantum waves superpose with definite phases, probabilities do not add everywhere. When quantum waves add with random phases, probabilities add everywhere. In Chapter 8 we will see that some physicists believe that the qualitative difference between random and coherent wave addition has important consequences for where one should draw the boundary line between quantum and ordinary reality.

FOURIER'S THEOREM

In 1798 Joseph Fourier, a talented French mathematician, accompanied Napoleon on his Egyptian adventure. He served for two years as governor of southern Egypt and in 1801 returned to France with a copy of the Rosetta stone. Examining the stone in Fourier's study, twelve-year-old Jean François Champollion was fascinated by its mysterious picture writing and vowed someday to translate it. Twenty years later Champollion achieved his goal and became the first person in three thousand years to read Egyptian hieroglyphics.

Champollion deciphered an ancient language which opened up Old Egypt to modern scholarship. Fourier, the man who showed the stone to Champollion, was also the discoverer of a new language, whose elements are not hieroglyphs but wave forms. Fourier's theorem, the key to the new wave language, is the foundation stone of all wave-based sciences including communications theory, modern optics, sound reproduction, oceanography, and quantum theory.

Fourier developed his waveform language to deal with heat waves. Heat as the motive power behind the burgeoning industrial revolution was an exciting mystery to nineteenth-century physicists and engineers. Lord Kelvin, the dean of English physicists, described *La Theorie analytique de la chaleur*, Fourier's elegant study of the flow of heat, as "a great mathematical poem." Fourier's theorem states that *any wave can be written as a unique sum of sine waves*.

The sine wave is a kind of undulatory archetype; its curvy profile is what most people have in mind when they visualize a wave. Vibrating strings and ripples in a pond are shaped each moment like sine waves. To see a sine wave standing still, look sidewise at a stretched spring (or any other helix).

Physicists like these waveforms because when they put a sine wave into any linear system, a similar sine wave always comes out. Linear systems change a sine wave's amplitude and phase but they never change its sinusoidal shape. Mathematicians like sine waves because no matter how many times they differentiate them, the result is always more sine waves. After listing the special attributes of this popular waveform, E. A. Guillemin celebrated the sine wave in words unusually colorful for an electrical engineer: "The sine wave is singled out as the one that shall forever be king and ruler." One can almost hear the fanfare of trumpets.

Imagine a wave w stretched out in space. Wave w is not necessarily oscillatory; it may take any shape whatever. Fourier's theorem says that wave w may be written as a sum of sine waves with various spatial frequencies k, amplitudes a, and phases p. Each word in Fourier's sine wave language is a sine wave with a different value for k, a, and p. Translating a wave into its sine wave words is called Fourier analysis. Wave w's Fourier analysis looks schematically like this:

$$w = (kap)1 \oplus (kap)2 \oplus (kap)3 \oplus (kap)4 \oplus \ldots$$

The particular sine waves which describe wave w are called its Fourier spectrum, or sometimes its vibration recipe. Each vibration recipe is unique: there is only one way to translate a wave into this sine wave language. The gist of Fourier's important discovery is that sine waves form a universal alphabet in terms of which any wave can be written.

Scientists have analyzed the sound waves produced by various musical instruments in terms of Fourier's sine wave alphabet. Even when sounding the same note, each instrument produces its own particular "tone color" —a difference reflected in its vibration recipe. Each instrument leaves a unique "Fourier fingerprint" in the air.

When we judge that a piano sounds different from a harpsichord, our brain may be attending not to different wave shapes but to different Fourier spectra. Physiological evidence (as well as our own experience) suggests that the human ear is sensitive to the sine wave content of sound. Coiled behind the eardrum, the cochlea, a tiny snail-shaped organ, turns sound into electrical impulses—brain code for auditory sensations. Each location along the cochlea's coil responds to a specific sine wave frequency. This little snail in the ear acts like a biological Fourier analyzer.

Just as a wave can be broken up into sine waves, so the same wave can be put together out of sine waves, an operation called "Fourier synthesis." Fourier's theorem tells us how to build any imaginable wave out of sine waves.

MUSIC SYNTHESIS

Until recently the sound of music was restricted to tonal qualities that could be produced by instruments which actually exist. Now Fourier's theorem provides the method and cheap transistor oscillators provide the means for the creation of entirely new tone colors—sounds impossible to produce by mechanical means. Composers of the New Music build sound directly out of sine waves from the keyboards of electronic synthesizers.

An electronic music synthesizer is a collection of sine wave oscillators whose amplitudes and frequencies can be varied in accord with customized vibration recipes. Each oscillator, vibrating at a selected amplitude, produces sine waves that are brought together in a mixer to create the desired tone color—either an imitation of an existing instrument or, more likely, some entirely novel electronic sound.

Electronic instruments based on Fourier synthesis are called analog synthesizers and were developed in the early sixties by New Music pioneers Robert Moog, Donald Buchla, and Paul Ketoff. The more recent digital synthesizers build their music not out of sine waves but out of waveforms called impulse waves. An impulse wave is an infinitely narrow spike of sound.

To picture digital analysis, imagine wave *w* going through a salami slicer. This machine breaks up a wave into very thin waveforms (impulse waves) which have the same amplitude as wave *w* at each location. As each slice of wave falls away, record its amplitude as a number on a list. These numbers are the wave's digital recipe or impulse spectrum. Plotting these numbers on a graph gives a curve that looks exactly like the original wave; the impulse spectrum of wave *w* is identical to the shape of wave *w* in space.

Digital synthesis involves looking at a wave's digital recipe and producing a string of narrow pulses whose amplitudes match the numbers on the list. Digital synthesis re-creates the salami by generating a sequence of slices the same size as the original.

Moog (analog) synthesizers typically have dozens of sine wave oscillators which are few enough to be tuned by hand. Digital synthesizers need to generate tens of thousands of "slices" to duplicate a sound lasting only a fraction of a second. The only practical way to handle so many numbers is in a computer memory. Analog synthesizers resemble a barbershop quartet—a handful of voices singing in chorus; digital synthesizers are full-fledged computers.

NEWTON'S PRISMS

Since 1822 when Fourier published his famous treatise on heat, sine wave analysis has developed into one of science's most valuable tools with thousands of practical applications. Recently Fourier techniques have proliferated due to development of computer programs which rapidly analyze complex waveforms into their sine wave components.

Wave analysis is rather new, as scientific discoveries go. However, more than a century before Fourier, Isaac Newton carried out a famous experiment which foreshadowed the French savant's new language of waves. By splitting sunlight with an upright prism into the familiar rainbow hues, Newton showed that white light is composed of colors. Newton named these colors "spectrum" from the Latin for "apparition." He also demonstrated that these colors could be recombined into white light by means of a second prism, inverted.

Borrowing from Newton, I symbolize any spectral analyzer—whether sine wave or impulse wave—by an upright prism. A spectral analyzer splits a wave up into component waveforms—sine waves in the case of a Fourier analyzer. I represent a typical output waveform by two parallel lines: a solid line standing for amplitude, a broken line representing phase, as a reminder that waves carry both these attributes. Although non-oscillatory waves have no cycles, time delay plays the role of phase for such waves. To complete the picture, I represent a spectral synthesizer by an inverted prism. Fig. 5.2 gives an example of this graphic convention.

THE SYNTHESIZER THEOREM

Fourier synthesis builds waves with a sine wave alphabet; digital synthesis creates the same waves out of impulse waveforms. If the same wave can be synthesized two different ways, why not more? Mathematicians attempting to extend Fourier's theorem in new directions made a remarkable discovery: almost any waveform family will work as the basic alphabet of a wave language.

This discovery—which I call the synthesizer theorem—means that

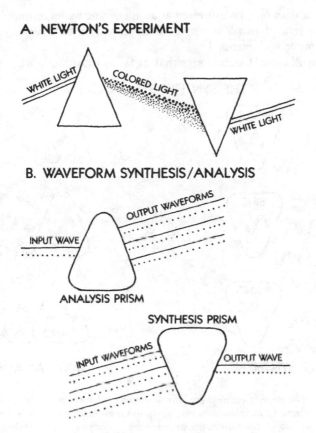

A. NEWTON'S EXPERIMENT

WHITE LIGHT

COLORED LIGHT

WHITE LIGHT

B. WAVEFORM SYNTHESIS/ANALYSIS

OUTPUT WAVEFORMS

INPUT WAVE

ANALYSIS PRISM

SYNTHESIS PRISM

INPUT WAVEFORMS

OUTPUT WAVE

FIG. 5.2 *A. Newton's separation of white light into colors was an early example of wave analysis. B. Following Newton, I use an upright prism to symbolize wave analysis into any waveform alphabet—not just sine waves. Prism input and output waves are represented by a double line: the solid line stands for wave amplitude, the dotted line for wave phase. An inverted prism symbolizes wave synthesis from members of any waveform family, not just the sine wave family. A machine that physically analyzes actual waves into a waveform alphabet is called a "hard prism"— symbolized by a triangle with sharp corners; a computer program that mathematically analyzes a waveshape into component waveforms is called a "soft prism"— symbolized by a triangle with rounded corners. Hard prisms output real waveforms; soft prisms give "vibration recipes." The art of breaking waves apart with soft waveform prisms is the heart of quantum theory.*

wave w can not only be expressed as a sum of sine waves or impulse waves but as a sum of piano waves, or flute waves, or tuba waves, or weird waveforms as yet unnamed.

The synthesizer theorem says that as far as providing a basic alphabet

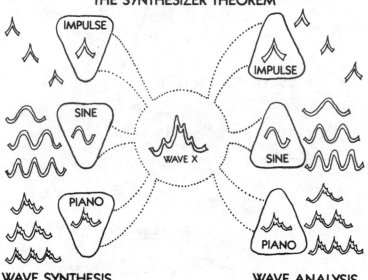

FIG. 5.3   *The synthesizer theorem: any waveshape whatsoever can be built (synthesis) from a sum of waveforms belonging to any waveform family—an innate wave property we might call "freedom of assembly"; likewise any waveshape can be decomposed (analysis) into a sum of waveforms belonging to any waveform family. The synthesizer theorem says that the same waveshape can be analyzed/synthesized in terms of a variety of different waveform alphabets: a wave, in short, has no intrinsic parts.*

for waves is concerned, there's nothing special about Fourier's sine waves. Any other waveform will do as well. By virtue of the synthesizer theorem a particular wave can be decomposed many different ways—as many ways as there are waveform families. This means that there is no "natural" way to take a wave apart. Unlike a clock, which breaks into gears and springs in just one way, a wave has no intrinsic parts.

WAVEFORM FAMILIES

Members of a human family share some common features, yet each is a unique individual. A member's family name tells what crowd he belongs to; a member's *personal* name picks him out from that crowd. It's the same with waveforms: each member of a waveform family has both a family name and a personal name.

Take for instance the *impulse wave family*. Each impulse wave looks exactly like any other—an infinitely narrow, spike-like wave. Impulse waves differ only in the place where they're located. An impulse wave's *position x* is its personal name, which distinguishes that wave from other members of the impulse wave family residing at other locations.

The *spatial sine wave family* consists of regular oscillations stretched out in space from horizon to horizon. All sine waves have the same shape but differ in spatial frequency. *Spatial frequency k* is a spatial sine wave's personal name.

The *temporal sine wave family* consists of regular vibrations in time which are distinguished from one another by their frequency *f*.

The *spherical harmonic waveform family* consists of the natural vibrations of a hollow sphere. The personal name of a spherical harmonic is made up of two integers, *n* and *m*, which distinguish one spherical harmonic from other members of its family. More information about this illustrious waveform family may be found at the end of this chapter.

KIN PRISM AND CONJUGATE PRISM

According to the synthesizer theorem, a wave *w* can be written in a waveform alphabet drawn from an infinite number of waveform families. However, for any particular wave there are two waveform families to which it bears a special relationship. There is one family to which it is *closest* in a certain sense, and one family from which it is *most distant*.

We analyze wave *w* into component waveforms by putting it through a waveform family prism, which separates any input wave into pure waveforms belonging to the prism's family. As we analyze wave *w* with different prisms we notice that some prisms break *w* into a few waveforms and some prisms break *w* into many waveforms. The number of waveforms into which a prism splits a wave is called that wave's spectral width, or sometimes bandwidth.

The size of this bandwidth bears an inverse relation to how closely wave *w* "resembles" the prism waveform which is analyzing it. The smaller the bandwidth, the more wave *w* resembles the prism waveforms; the larger the bandwidth, the less the family resemblance. For instance, if we put wave *w* into its own family prism (call this the W family), it will be analyzed into only one waveform, namely itself—the minimum possible bandwidth. I call this prism—the prism that does not split wave *w* at all—its kin prism. The waveforms associated with this prism are its own kind, the family to which it belongs.

On the other hand, among all the waveform families in the world, there is one family (family M) whose prism gives the largest possible bandwidth when it's used to analyze wave *w*. The members of family M resemble *w* the least. I call this prism—the prism that splits wave *w* the most—its conjugate prism.

One can imagine every waveform family inhabiting a spherical wave-space like the Earth's surface. Families which resemble the W family

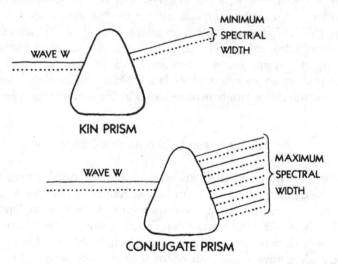

FIG. 5.4   *Kin prism and conjugate prism. A wave's kin prism is that analysis prism which splits it the least; a wave's conjugate prism is that analysis prism which splits it the most. A wave belongs to its kin prism's waveform family and resembles least the members of its conjugate waveform family.*

live nearby; families which are different from W live farther off. The family most distant from W is its conjugate family M, which lives at the antipodes—precisely on the other side of the world.

Just as every wave belongs to a unique waveform family, so every waveform family possesses a conjugate family whose members are its polar opposites.

The sine wave family (the basis for analog music synthesis) consists of smoothly oscillating waves without beginning or end. The sine wave's conjugate waveform is the impulse wave (the basis for digital synthesis). An impulse wave is a single narrow spike that lasts for an immeasurably small instant. You could hardly imagine two waves more different than these conjugate waveforms, sine wave and impulse wave.

SPECTRAL AREA CODE

Consider a pair of conjugate waveform families, W and M, which necessarily dwell in opposite regions of their spherical wavespace. We now imagine putting an arbitrary wave X into each of these polar opposite family's prisms.

When we put X into the W prism we get a particular bandwidth ΔW of output W waveforms; this bandwidth is an inverse measure of how closely the input wave X resembles the members of the W waveform family. When we put X into the M prism we get a particular bandwidth ΔM of output M waveforms; this bandwidth is an inverse measure of how closely the input wave X resembles the M waveform family.

Since W and M are polar opposites, it's impossible for wave X to resemble them both no matter how contorted its wave shape. Consequently there must be some restriction on how small these two spectral widths can get for the same input wave. Indeed, a rule exists which expresses the impossibility of making both these spectral widths small with the same wave. This rule looks like this:

$$\Delta W \Delta M > 1$$

For any wave X, the product of its W spectral width and its M spectral width—where W and M are conjugate waveform families—can never be smaller than 1.

Since the product of two spectral widths is a kind of spectral area, the content of this rule is that for any wave X, X's spectral area must exceed one unit of spectral space. If you want to make a wave—no matter what its shape—you will have to provide a certain minimum spectral acreage to put it on. Since this rule resembles a kind of building code in spectral space, I call it the spectral area code. An example of the spectral area code

THE SPECTRAL AREA CODE

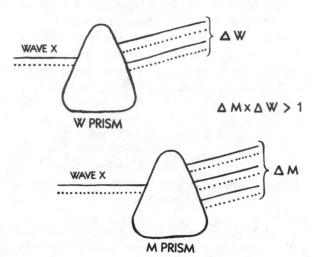

$$\Delta M \times \Delta W > 1$$

FIG. 5.5   *The spectral area code. No wave X can resemble both waveform W and its conjugate waveform M: the more it resembles W, the less it looks like M. In terms of spectral widths, this logical constraint requires that the product of a wave's W width and its M width—what might be called its spectral area—can never be smaller than a certain constant, fixed by nature for all waves.*

in action is the complementary abilities of analog and digital synthesis techniques.

An analog synthesizer builds a sound wave X out of a range of sine waves with different spatial frequencies k. Each wave X, depending on its shape, will require a certain spectral width Δk of sine waveforms for its proper analog synthesis.

The sine wave's conjugate waveform is the impulse wave, the basis of digital music synthesis. A digital synthesizer builds a wave X out of a range of impulse waves with different values of position x. Each wave requires a certain spectral width Δx of impulse waves for its proper digital synthesis.

For any waveform X, the spectral area code requires mutual constraints on X's analog and digital bandwidths:

$\Delta k \Delta x > 1$ (spectral area code for analog/digital synthesis)

Short percussive musical sounds, such as castanets, triangle, woodblock, have a narrow impulse spectrum since their waveforms take up little space. To analog-synthesize such crisp sounds adequately, the spectral area code requires a large range of sine waves. To synthesize an infinitely short sound, the impulse wave itself, would require all possible sine wave frequencies.

On the other hand, musical sounds that are nearly pure tones, such as flute, organ, tuning fork, have a narrow sine spectrum. To digitally synthesize such pure tones adequately, the spectral area code demands a large range of impulse waves.

The spectral area code tells us that analog and digital music synthesizers are complementary in a particular sense: one is good for synthesizing long waveshapes, the other for short ones. The spectral area code is a basic feature of all waves, as inseparable from their undulatory nature as diffraction and interference. This code is important in optics, where it limits the resolving power of microscopes, in communications theory, where it fixes the bandwidth of TV channels, and in numerous other wave-based operations. As we shall see, the spectral area code when applied to waves of quantumstuff leads immediately to the Heisenberg uncertainty principle.

Although in its application to ordinary waves this code has been known for more than a century, it never received a proper name. Now due to semantic backformation more and more textbooks on ordinary wave theory refer to this natural limitation on mutual spectral widths as the uncertainty principle despite the fact that in its application to ordinary waves it has nothing to do with uncertainty.

Whenever it is measured the world seems solid (particle-like), but the pattern formed by these particles leads to the conclusion that between measurements the world acts like a wave. Following this lead, quantum theory represents the unmeasured world as a wave identical in its behavior to ordinary waves, but interpreted in a decidedly non-ordinary manner. This brief survey of basic wave behavior is intended to provide solid ground from which we can venture into the slippery territory of quantum interpretation.

EPILOGUE: A FAMILY OF SPHERICAL WAVES

Looking at droplets from a leaky faucet in the flickering glare of a strobe light, you seem to see hanging in midair single quivering balls of water. The oscillation of water droplets is conveniently described in terms of a waveform family called the spherical harmonics.

Just as sine waves are the natural vibrations of a stretched string, so spherical harmonics are the natural vibrations of an elastic sphere. If it is shaped like a ball, chances are that some scientist has described it in the spherical harmonic wave alphabet. Members of this waveform family portray the figure of the Earth and its magnetic field, the radiation pattern surrounding a TV antenna, oscillations of raindrops and atomic nuclei, the heave of ocean tides, and the agitation of numerous spherical resonators, from temple gongs to the sizzling surface of the sun.

Besides its family name, each spherical harmonic has two personal names which distinguish that member's waveshape from the rest of the family. Whenever a sphere vibrates, certain *nodal circles* appear where the sphere stands still. On one side of the nodal circle, the sphere's surface is moving in; on the other side of this boundary, the sphere is moving out.

A spherical harmonic's first name (commonly called its "order") is a number n which counts its total number of nodal circles. The simplest spherical harmonic has no nodal lines at all *($n = 0$):* the entire sphere expands and contracts as a whole, a style of spherical motion called the "breathing mode." Each spherical harmonic possesses an axis of symmetry marked like the axis of the Earth by north and south poles. A spherical harmonic's second name (commonly called its "degree") is a number m which counts how many of its nodal circles pass through its poles. A general property of spherical vibration is that if a nodal circle doesn't go through the poles, then it must lie in a plane parallel to the sphere's equator. Consequently all nodal circles are lines of definite latitude or longitude.

For instance the spherical harmonic labeled $n = 5$, $m = 4$ (fifth order, fourth degree) has five nodal circles in all, four of which pass through its poles. The odd nodal circle must be a line of latitude which, for reasons of symmetry, passes through the equator. For the spherical harmonic waveform family, two numbers suffice to list all family members. Nobody is left out.

SPHERICAL HARMONIC WAVEFORMS

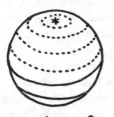

n = 1, m = 0

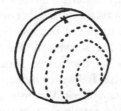

n = 1, m = 1

n = 3, m = 2

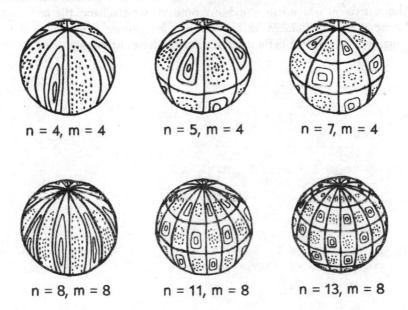

n = 4, m = 4

n = 5, m = 4

n = 7, m = 4

n = 8, m = 8

n = 11, m = 8

n = 13, m = 8

FIG. 5.6 *Spherical harmonic waveform family. A member of this celebrated family —the natural vibrations of an elastic sphere—is distinguished by a number n, which counts its nodal circles, and a number m, counting those nodal circles that go through the sphere's poles.*

Next to the sine waves, no waveform family is more highly regarded by scientists working with waves than the spherical harmonics. By virtue of the synthesizer theorem, any spherical vibration, no matter how complicated, can be expressed in the spherical harmonic waveform alphabet. The behavior of these useful waveforms has been studied for more than a century. That the personal names of the spherical harmonics are integers is of particular interest for the purpose of quantum theory.

All waveform families so far considered possess a continuous range of names—quantities like spatial frequency, which can take on any possible numerical value. In contrast, spherical harmonics possess personal names which are discrete. Waveform families with discrete names are like fretted instruments (guitar, trumpet, piano) which can play only a limited number of notes, compared to the unfretted instruments (violin, trombone, human voice) which can sound any note. As we shall see, the fact that some waveform families have discrete names means, when applied to quantum waves, that certain physical attributes must be quantized.

6

Meet the Champ: Quantum Theory Itself

> By nature I am peacefully inclined and reject all doubtful adventures. But a theoretical interpretation had to be found at all costs, no matter how high . . . I was ready to sacrifice every one of my previous convictions about physical laws.
>
> Max Planck.

A visitor to Niels Bohr's country cottage asked him about a horseshoe nailed above the front door. "Surely, Professor Bohr, you do not really believe that a horseshoe over the entrance to a home brings good luck?" "No," answered Bohr, "I certainly do not believe in this superstition. But you know," he added, "they say it brings luck even if you don't believe in it."

Quantum theory is like Bohr's horseshoe: it works no matter what a person believes. One quantum theorist may imagine she's charting the destinies of multiple worlds; another fancies he's thinking quantum logically. Despite different notions about what they're doing (indicative of physicists' confusion about what quantum theory actually means), both theorists will come up with the same result.

Quantum theory was devised in the late twenties to deal with the atom, a tiny entity a thousand times smaller than the wavelength of green light. Disturbed by its philosophical implications, many physicists at that time considered quantum theory a provisional device bound to fail outside the atomic realm. Quantum theory continued, however, to prosper beyond its inventors' wildest dreams, resolving subtle problems of atomic structure, tackling the nucleus some ten thousand times smaller than the atom itself, and then extending its reach to the realm of elementary particles (quarks, gluons, leptons) which many believe to be the world's ultimate constituents. With each success quantum theory became more audacious. Quantum physicists looking for new worlds to conquer turned their sights to the macrocosm, and now dare to model the birth of the universe itself as one gigantic quantum jump. Heaping success upon success, quantum theory boldly exposes itself to potential falsification on a thousand different fronts. Its record is impressive: quantum theory passes every test we can devise. After sixty years of play, this theory is still batting a thousand.

Before looking at the quantum reality question, which sharply divides them, let's begin where physicists all agree: how to actually use quantum theory. From Berkeley to Gorky, quantum physicists predict quantum facts in exactly the same way. In this chapter I treat quantum theory strictly as a tool for predicting experimental results, and do not inquire at all as to what it might mean.

Considered merely as a tool, quantum theory is a conceptual recipe which predicts for any quantum entity which values of its physical attributes will be observed in a particular measurement situation. Quantum theory by design only predicts the results of measurements; it does not tell us what goes on in between measurements.

Quantum theory predicts the results of measurement with unsurpassed accuracy, but measurements are only part of the world. Most everywhere, most of the time, the world dwells in an unmeasured state. Anyone curious about reality will want to know what the world is like when it is not being measured. Quantum theory does not directly address this question.

We can get a sense of how quantum theory operates by answering three questions:

1. How does quantum theory describe a quantum entity?
2. How does quantum theory describe a physical attribute?
3. How does quantum theory describe a measurement situation?

Once we understand how physicists actually use quantum theory to predict a quantum entity's measured attributes, we will look at some attempts to go beyond quantum theory to the unmeasured world itself.

QUANTUM ENTITIES

Although the world once appeared to be twofold, made of particles and fields, closer observation reveals a common behavior. Former "particles" now show their wave aspects; former "waves" behave like particles. In reality everything is made of the same kind of substance, which I call quantumstuff. Quantum theory reflects this fundamental unity of being by describing all quons the same way. One description fits all.

Step one in quantum theory is to associate with each quantum entity a particular wave called that quon's wave function, customarily labeled ψ or *psi*, the twenty-third letter in the Greek alphabet. Most physicists treat the wave function as a mere calculational device, not as a real wave located somewhere in space. Just as the President and members of Congress represent individual citizens in a very particular but indirect manner, so a wave function likewise represents individual quons in a specific but indirect way —a way which will become clearer when we see how a wave function is actually used. Meanwhile we may imagine each quon—an electron in the Airy experiment, for instance—to be represented by its own proxy wave bearing some as yet unspecified relation to the quon itself.

Like any other wave, a quon's proxy wave enjoys all the benefits of wave nature outlined in the previous chapter. In particular, a quon's wave function possesses amplitude and phase, satisfies the superposition principle and spectral area code, and displays phase-dependent constructive and destructive interference. Most important for quantum theory, the wave function by virtue of the synthesizer theorem can be expressed as the sum of members of numerous waveform families.

One way in which ψ differs from ordinary waves is that it carries no energy. For an ordinary wave, the square of its amplitude measures its energy. For a quantum wave, the square of its amplitude (at location x) represents not energy but probability, the probability that a particle—a localized packet of energy—will be observed if a detector is placed at location x. Because the quantum wave carries no energy, it is not directly detectable; we never see any quantum waves, only quantum particles. However, after many particles have appeared we can infer the presence and the shape of the ψ wave from the pattern of particle events. In Fig. 4.2, for instance, phosphor flashes trace the shape of the electron's ψ wave.

Since ψ^2 stands for probability, the wave function is often called a probability wave. However, it is ψ, not ψ^2 which actually represents a quantum entity. I respect this difference between the wave function and its square by calling the quantum proxy wave a possibility wave—a possibility being somewhat less real than a probability. We shall see later how appropriate the name "possibility" is for this kind of wave. The relationship between quantum possibility and probability is simple:

$$probability = (possibility)^2$$

The amplitude of a quantum wave is its possibility. The square of a possibility is a probability.

More particles will be detected where ψ^2 is large; fewer where ψ^2 is small. The pattern of detected quons—flashes on a phosphor screen, for instance—allows us to infer the shape of the probability wave ψ^2 and hence the shape of the possibility wave ψ. The location of phosphor flashes makes ψ indirectly visible, just as smoke particles outline the shape of invisible air currents. Because this wave acts behind the scenes, so to speak, carrying no energy and revealing itself only indirectly through its statistical influence on a large number of particle events, Einstein called ψ a *Gespensterfeld* or ghost field. Since it carries no energy, the wave function is also referred to as an empty wave. In France, the ψ wave is called by a beautiful name—*densité de présence,* or "presence density."

Probability is a measure of the relative number of ways a particular event can happen. The science of probability began as a collaboration between mathematicians and gamblers to determine the odds in dice games by systematically counting the ways a desired outcome could occur. Probability has long outgrown its shady past and now plays an important role in economics, industry, and scientific research, but many of its applications still involve little more than complicated counting schemes.

Since it is based on counting things, probability obeys the rules of ordinary arithmetic. For instance, if an event can happen five ways and three new ways open up, the event then happens eight ways. This common-sense addition rule, a central feature of classical probability, is not always respected by quantum probabilities, which follow the laws of wave addition rather than ordinary arithmetic.

Interference between waves is not unusual; it happens every day in your kitchen sink. Interference of quantum waves, however, gives peculiar results. Because quantum possibilities add like waves, not like things, physical possibilities can vanish if their representative waves happen to meet out of phase.

Not long ago a simple case of interfering possibilities changed the course of elementary-particle physics. In the late sixties physicists had good reason to believe that all strongly interacting particles (hadrons) were actually composed of a few truly fundamental particles called quarks. At that time, combinations of only three quarks—distinguished by a "flavor" quantum number as up, down, or strange—sufficed to account for the existence of dozens of known hadrons.

In 1970 Harvard physicist Sheldon Glashow conjectured that a fourth (charm-flavored) quark must exist and he estimated both the properties of the charmed quark and new yet-to-be-discovered hadrons that could be built from it. One of Glashow's best arguments for the charmed quark's existence was the nonexistence of a particular decay mode of the K-particle. Like most of its sister hadrons, the K-particle is unstable and decays in numerous ways into less massive particles. Yet despite long searches, the K-particle had never been observed to decay into two muons.

According to the three-quark model of hadrons, the K-particle's quark structure demanded that it decay into a pair of muons in a certain way— via the strange quark channel. (The up and down quarks can't make muons.) Yet this decay mode never seemed to occur. Would physicists have to scrap the quark model because it predicted a process that didn't happen?

Glashow's plan for eliminating two muon decay was not to do away with quarks but to add one more. He postulated a second way that the K-particle could change into two muons, a way which involved a new quark—the charmed quark channel. In classical physics, if you double the number of ways something can happen, you expect that it will happen twice as often. Not so in quantum physics, which permits the cancellation of possibilities. Glashow tailored the properties of the new quark so that its possibility wave was opposite in phase from the strange quark's wave. The two possibilities cancel, nicely suppressing two-muon decay. Actually, because these quarks have different masses the cancellation is not quite exact. Recent measurements of K-particle decay have uncovered a few double-muon events, less by a factor of a million from the three-quark predictions but in line with Glashow's four-quark model. So convinced was Glashow that the absence of muons signaled the presence of charm,

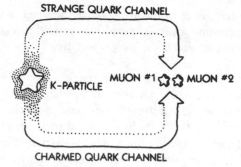

FIG. 6.1   *More is less. The K-particle can decay into a muon pair in two ways—via the strange-quark and the charmed-quark channels. However, because these two possibilities have opposite phases, they cancel in the peculiar quantum fashion which suppresses the occurrence of the double-muon decay.*

that at the Fourth International Conference on Meson Spectroscopy (April 1974) he promised to eat his hat at their next meeting two years hence if charm was not discovered in the meantime.

Glashow did not have to eat his hat. In November 1974 the J-particle was discovered, the first of the new charm-constituted hadrons. Subsequently other charmed particles showed up. Today quark theory is a cornerstone of particle physics. Theorists see a need for two more quark flavors (top and bottom), making six flavors in all. Confidence in the existence of quarks is now so high that, unlike the suspense attending Glashow's bold conjecture, confirmation of top and bottom quarks is regarded as almost inevitable.

This destructive interference of a quark's physical attribute (its ability to decay into muons) represents a routine application of quantum concepts at the frontiers of present knowledge. This example reinforces our belief that quantum theory applies to all physical entities without exception. No entity is so exotic that it escapes the quantum rules: a quark's a quon too. One description fits all.

The answer to our first question—How does quantum theory describe a quantum entity?—is this: quantum theory does not "describe" entities at all; it represents them. Instead of dealing directly with a quantum entity, quantum theory replaces it with a proxy wave ψ, whose square at any location gives the probability that the quon's particle aspect will manifest there (position attribute) and whose shape gives information about all attributes other than position, in a manner to be described next.

QUANTUM ATTRIBUTES

The most significant word in the quantum vocabulary is the term "attribute." Quantum theory's bizarre nature stems primarily from the unusual way it represents physical attributes.

Each quon possesses two kinds of attributes: static and dynamic. A static attribute always has the same size each time it is measured, and this serves to distinguish one type of quon from another. The most important static attributes are mass *(M)*, charge *(Q)*, and spin magnitude *(S)*. Today there are more than a hundred elementary quons, but when quantum theory began only three were known: electron, proton and photon. The sizes of the static attributes of these classic quantum entities are as follows:

QUON	MASS	CHARGE	SPIN
Electron	·1	− 1	1/2
Proton	1836	+ 1	1/2
Photon	0	0	1

Static attributes are fixed for life, but a quon's dynamic attributes change with time. The major dynamic attributes are position, momentum, and spin orientation, which tell where a quon is, how it's moving, and in what direction its spin is pointing.

Position attribute keeps track of a quon's changing location. For a point quon like electron, position labels the whereabouts of that point. For an extended quon like an atom, position refers to the quon's center of mass. Since space has three fundamental directions, conventionally entitled x, y, z, position attribute also consists of three parts, X, Y, Z, which specify a quon's location anywhere in three-dimensional space.

Because M already stands for mass, the dynamic attribute linear momentum (often shortened to momentum) is symbolized by the letter P. A quon's momentum measures how fast it is moving and in what direction. Like position, momentum has three components, symbolized P_x, P_y, P_z, which tell how fast the quon's moving in the x, y, z directions.

Classically a particle's spin attribute (also called intrinsic angular momentum) is proportional to the particle's mass and its angular velocity (how fast it is turning around). Like linear momentum, spin points in three directions. Physicists define an entity's spin to lie along its axis of rotation, pointing up or down depending on the sense of rotation *(CW* or *CCW)*. According to this convention, the Earth's spin looks like an arrow parallel to its axis which points out of the North Pole.

Spin is an arrow that can point in any direction: the symbols S_x, S_y, S_z label the components of the spin orientation attribute in the x, y, z directions. A convenient dimensionless way of speaking about spin is to specify what fraction of the total spin points in each direction. For instance the spin may point 50 percent in the x direction and 25 percent in each of the y and z directions.

The Earth's spin, for instance, is inclined to the ecliptic axis (the axis of the Earth's motion around the sun) by about 23 degrees, a circumstance that gives rise to the seasons. For this 23-degree tilt the fraction of the Earth's spin that points along the ecliptic axis is 85 percent. If we designate this axis by z, the Earth's spin orientation attribute S_z is equal to 0.85. The remaining 15 percent of the Earth's spin is divided between the S_x and S_y attributes.

In addition to spin orientation, a body possesses spin magnitude, *S*. A heavy, rapidly turning object like the Earth has a large spin magnitude; a light, slowly rotating object like a snowflake has a small spin magnitude.

Unlike the Earth and the snowflake, which can have various spin magnitudes, an elementary quon cannot change its spin magnitude; for all elementary quons, *S* is a static attribute. This means, for example, that electrons can never increase or decrease their spin; they rotate eternally at the same rate. The spin orientation of elementary quons, on the other hand, can take different values. For all entities, quantum or classical, spin orientation is a dynamic attribute.

Incidentally, physicists often use the word "spin" for both the spin magnitude and the spin orientation attributes. Whether "spin" refers to a quon's static or dynamic attribute is usually clear from the context.

Quantum theory represents every dynamic attribute in the same unusual way. One description fits all. This theory associates each dynamic attribute with a particular waveform family. Likewise every conceivable waveform family corresponds to some dynamic attribute. Within a particular waveform family, different family members stand for different physical values of that family's adopted attribute. The secret of how quantum theory works lies in its association of waveforms with attributes. A few examples will make this curious waveform-attribute connection more concrete. The position attribute is associated with the impulse wave family of waveforms. Each impulse wave represents a different value of position. A typical member of this family is an extremely tall, infinitely narrow pulse located exactly at position x. The position value X associated with this family member is the same position x at which this waveform is located:

$$X = x$$

The momentum attribute is associated with the spatial sine wave family of waveforms. Each sine wave represents a different value of momentum. A typical member of this family resembles an infinitely long oscillation stretching into the distance with spatial frequency k. The momentum value P associated with each family member is given by the relation:

$$P = hk$$

where h is Planck's constant of action.

The energy attribute is associated with the temporal sine wave family of waveforms. Each sine wave represents a different value of energy. A typical member of this family is a pure musical note vibrating at frequency f. The energy value E associated with each family member is given by the relation:

$$E = hf$$

where h is again Planck's constant

The spin attribute is associated with the spherical harmonic family of waveforms. Each spherical harmonic represents a different value of spin magnitude and spin orientation. A typical spherical harmonic looks like a globe of quivering jelly whose regions of vibration are partitioned by a certain number n of nodal circles, a number m of which pass through the poles of the sphere. The spin magnitude S is given by the total number of nodes n according to the rule:

$$S = hn$$

where h is Planck's constant. The spin orientation depends on the number of nodes m that go through the poles: the more polar nodes, the more the spin points in the polar direction which we will call z. For a particular spherical harmonic, the spin orientation value S_z in the polar direction is given by the relation:

$$S_z = m^2/(n^2 + n)$$

WAVEFORM–ATTRIBUTE DICTIONARY

WAVEFORM   ATTRIBUTE

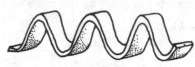

IMPULSE WAVE

POSITION

SINE WAVE

MOMENTUM

SPHERICAL HARMONIC

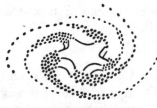

SPIN

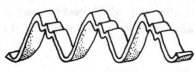

PIANO WAVE

UNNAMED ATTRIBUTE

FIG. 6.2 *Quantum waveform-attribute dictionary. Every waveform family corresponds in quantum theory to a physical attribute—a universal quantum code which is the key to much of this theory's peculiar behavior. Along with each waveform-attribute entry goes a rule which connects a wave quantity (personal name) with the magnitude of its corresponding physical attribute.*

Because n and m count nodal circles, they are restricted to integer values. This means that both spin magnitude and spin orientation can take only certain discrete values. Such attributes are said to be "quantized."

Classical physics placed no limits on the values an object's spin could take: Newton's apple could spin any way it pleased. Quantum theory, on the other hand, requires spin and other special attributes to be quantized —they can take only certain values and not others. Quantized attributes are digitized: they can be measured with perfect accuracy, a quantum fact which compensates somewhat for the fundamental quantum inaccuracy required by Heisenberg's uncertainty principle.

Quantum theory's waveform-attribute connection explains why some attributes are quantized and others are not. Quantized attributes correspond to confined waveforms like the spherical harmonics, whose vibrations are restricted to the surface of a sphere.

An unconfined wave—a sine or impulse wave, for instance—vibrates as it pleases; a confined wave vibrates only at certain resonant frequencies. This constraint on vibratory states translates into a constraint on numbers of waveform family members that explains why not all values of certain attributes can occur in nature.

Quantum theory's association of attributes with waveforms can be expressed as a quantum waveform-attribute dictionary listing for every dynamic attribute its corresponding waveform family. A portion of this quantum dictionary is shown in Fig. 6.2. Along with each waveform-attribute entry in the quantum dictionary goes a rule for translating a waveform's personal name into the size of its corresponding attribute. This rule usually (but not always) involves Planck's constant.

Here is a list of some of these rules:

WAVEFORM FAMILY NAME	WAVEFORM PERSONAL NAME	DYNAMIC ATTRIBUTE	ATTRIBUTE SIZE
Impulse	x	Position	$X = x$
Spatial sine	k	Momentum	$P = hk$
Temporal sine	f	Energy	$E = hf$
Spherical harmonic	n	Spin mag.	$S = hn$
Spherical harmonic	n, m	Spin orient.	$S_x = m^2/(n^2 + n)$

We recognize in this list the formulas $P = hk$ and $E = hf$. These are just de Broglie's law for the electron's wavelength and the Planck-Einstein relation for the energy of a quantum of light. Quantum theory explains these historic equations as simple consequences of the waveform-attribute connection. But these two rules are just the tip of the quantum iceberg: for every dynamic attribute there is such a formula which connects the size of some mechanical attribute with a particular wave property: the personal name of its corresponding waveform.

The waveform-attribute connection is without a doubt one of quantum theory's most unusual features. It seems natural to associate a wave with each quon since quons show some wavelike aspects. But how can we justify the association of abstract waveforms with mechanical attributes? What on earth does momentum have to do with sine waves?

Ultimately this waveform-attribute association is justified because it works. A realist would say that it works because it reflects some correspondence really present in the world: nature has associated momentum with sine waves from the beginning. Humans have only recently discovered the naturalness of this connection. For thousands of years our culture has been shaped by literature, liturgy and legislation built of human language. In *The Cosmic Code* Heinz Pagels pictures the scientific enterprise as the opening of the "Great Book of Nature": the discovery and decoding of the ancient non-human message which orders the universe. The cosmic code entered our awareness only yesterday, but already the strange beauty of this alien language is restructuring human culture to its own design.

The biology sector of the cosmic code is dominated by the DNA code, which associates certain molecular trigrams with particular amino acids—the building blocks of life on Earth. The key to the cosmic code's quantum sector is a code that assigns a particular waveform to every physical attribute. The quantum waveform-attribute code is more general than DNA: sinewave means momentum for quons all over the universe.

According to quantum theory any waveform, no matter how bizarre, corresponds to some particular dynamic attribute which we could in principle measure. There is an infinity of possible waveform families, which means that the waveform-attribute dictionary contains an infinite number of entries.

For instance the "piano" waveform connects to some presently unknown mechanical attribute—call it the piano attribute—which an electron or any other quon is bound to display in a piano measurement situation. Likewise we could test an electron for the size of its tuba attribute, its flute attribute, or its Wurlitzer organ attribute. Physicists have shown little interest in measuring such obscure mechanical properties, but should the need ever arise quantum theory can predict these results as easily as it predicts the results of spin and momentum measurements.

The answer to our second question—How does quantum theory describe a physical attribute?—is this: quantum theory does not "describe" attributes; it represents them. Instead of dealing directly with a quon's mechanical attributes, quantum theory replaces each one with a particular waveform, drawn from a universal quantum waveform-attribute dictionary. Once we've represented our quon (with a proxy wave ψ) and the attributes (waveforms) we intend to measure, we are ready to tackle the measurement act itself.

QUANTUM MEASUREMENT

To describe a measurement situation, quantum theory relies on the synthesizer theorem, which states that any wave w can be broken up into waveforms belonging to any waveform family. This theorem allows us to imagine an analyzer prism for each waveform family: a sine wave prism, for instance. The sine wave prism works as follows: you put wave w into the prism and it breaks this wave into a spectrum of output sine waves.

The prisms we have in mind here are "soft prisms," not physical objects. They stand for the mathematical analysis of an input wave into a sum of waveforms (vibration recipe) which belong to a particular waveform family. Following Joseph Fourier's lead, mathematicians for more than a century have been devising clever ways to break waves up into component waveforms. Their efforts have been translated into computer programs which quickly analyze an arbitrary input wave into any of a number of common waveform families. When we say "sine wave prism," we should imagine one of these computer programs separating a wave into its unique sine wave constituents.

Let's suppose that we possess a box of such soft prisms (programs), one prism for each waveform family. Now we have on hand everything we need to describe a quantum measurement. Suppose we've decided to measure an electron's "A-attribute" where A may be position, momentum, or any other mechanical attribute no matter how exotic. Maxine, the clever theoretical physicist, will predict the outcome of this experiment using quantum theory. Let's peek over Maxine's shoulder and watch how she uses quantum theory to describe the measurement of attribute A.

Step one is to determine the electron's wave function ψ. A large part of a physicist's education consists of learning how to construct the wave functions of quons that find themselves in drastic physical situations, such as being locked in boxes, dropped into deep wells, or pushed through tiny orifices. We suppose such training has done its job and Maxine knows how to calculate for the electron in question the shape of its quantum proxy wave.

Step two is to look up attribute A in the waveform-attribute dictionary. Suppose Maxine finds that attribute A corresponds to waveform family W. The members of the W waveform family are distinguished by a wave property w which uniquely identifies each member. The dictionary also provides a rule that associates each wave property w with an A-attribute value A. This rule may be as simple as:

$$A = hw$$

or it may be quite complex. In any case, given waveform $w1$, Maxine knows the value $A1$ of its associated physical attribute. We assume for simplicity that A is a quantized attribute, which takes only two possible values, $A1$ and $A2$. Hence there are only two waves in the W waveform family, one wave labeled $w1$ and the other labeled $w2$.

An attribute which takes only two values is the simplest possible kind of dynamic attribute. However, despite their simplicity two-valued attributes display the full range of quantum behavior and play an important role in physics: both the spin orientation of the electron and the polarization attribute of the photon take only two values.

Step three: Maxine reaches into her box of soft analyzer prisms and pulls out the prism which breaks waves into W waveforms.

Step four: Maxine puts the electron's proxy wave ψ into the W prism and notes what comes out. The W prism divides the electron's proxy wave into two waveforms, $w1$ and $w2$, each with its own amplitude and phase. We recall that for all quantum waves, intensity (amplitude squared) stands for probability.

Step five is to square the amplitude of each waveform to get two probabilities. Suppose these numbers are 0.75 (75 percent) and 0.25 (25 percent). This step concludes the quantum description of the measurement process. We are now ready to test quantum theory's predictions against the quantum facts.

To compare quantum theory with fact, we must find a way to measure attribute A on the electrons in question. This means constructing in the laboratory an A-attribute analyzer, which is not a mathematical procedure like Maxine's prisms but an actual material object. The A-analyzer is a machine that works like this: you put in a beam of quons (each represented by the same wave function) and the A-analyzer sorts them into two beamlets, one whose quons all have A value $A1$, and the other whose quons have A value $A2$. In each of these beamlets sits a detector (which could be a phosphor screen) which counts and records the number of quons in each channel. To measure any attribute, simply count how many quons display each value of that attribute.

Quantum theory, in this case, predicts that 75 percent of the electrons will show value $A1$ and the remaining 25 percent will end up in the $A2$ detector. In every case where probabilities like these have actually been compared with experiment, the facts have agreed with theory. Had this been an actual test, with any quon beam, with any attribute, no physicist would wager against quantum theory. In millions of experimental situations, in hundreds of different laboratories, this unusual predictive procedure has never missed the mark. In the lab, an experimentalist deploys an actual quon beam, an actual attribute sorter which separates these quons into beamlets with different attribute values, plus actual detectors which count the number of quons in each beamlet.

On paper, a theorist replaces the quon beam with a fictitious proxy wave and replaces the attribute sorter with a soft waveform analyzer, which breaks the quon's proxy wave into a spectrum of fictitious waveforms which take the place of attributes in this theory. The intensity (amplitude squared) of each output waveform predicts the probability that its corresponding attribute will appear in the experimental attribute spectrum.

Quantum theory parallels the actual measurement of a quon's physical attributes with a mathematical charade starring a fictitious proxy wave and its constituent waveforms. This wholly wavelike representation is reconciled with the fact that every actual quantum event is (like the flash on a phosphor screen) wholly particle-like by construing these quantum waves to mean not the events themselves but the pattern of these events. Because they correspond to patterns of particles, these quantum proxy waves behave less like ocean waves, more like "crime waves"—the kind of wave that governs the shape of the forest rather than the location of individual trees.

In practice quantum theory cannot be distinguished from classical theory in many situations (quantum gravity, for instance) because the new

A. QUANTUM REPRESENTATION

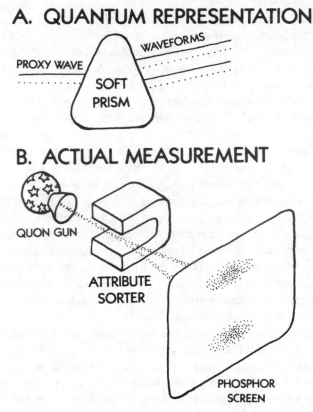

B. ACTUAL MEASUREMENT

FIG. 6.3   *Representation of a quantum measurement. Quantum theory represents a physical measurement as a soft prism which analyzes a proxy wave into component waveforms. This conceptual procedure stands for an experiment in which a piece of laboratory hardware sorts quons into different beams depending on the size of a particular physical attribute.*

effects it predicts are too tiny to measure. In addition the scope of both quantum experiment and theory is limited by the experimental and computational means of present-day technology. One of these technical limitations is the difficulty of computing the waveforms which correspond to certain attributes. Theorists can easily calculate the waveform-attribute connection for attributes of elementary quons, but these calculations rapidly become more difficult as the quons get more complex. The calculation of the attributes of molecules is not easy; to compute the waveform which corresponds to the heads/tails attribute of a flipped penny is beyond the reach of today's mathematics.

On the experimental side, it's easy to measure flipped pennies, but often very difficult (and expensive) to construct for elementary quons the attribute sorters we so casually invoke on paper. On the other hand, for an elementary quon's major attributes, physicists have been able to devise such sorters.

The position attribute can be measured with a phosphor screen or with several other detectors sensitive to a quon's raw presence. Your eye is a position-attribute sorter of photons.

For charged particles a uniform magnetic field acts as a momentum sorter, and for neutral particles a non-uniform magnetic field can sort quons according to their spin magnitudes and orientations. As we shall see, certain transparent crystals can sort photons according to their polarization attribute, an attribute of light closely related to the photon's spin.

Some attribute sorters cost millions of dollars—the machine, for example, which counts how many K-particles possess the two-muon decay attribute; other sorters are extremely simple. The Airy experiment (Fig. 4.1) is an example of a momentum sorter consisting of nothing but a small hole and a phosphor screen. Any quon which passes through the hole with zero sidewise momentum strikes the center of the screen. The more (sidewise) momentum a quon has, the more it will drift sidewise in transit and hit the screen off center. The Airy pattern, in addition to demonstrating the wave nature of electrons, also records the (sidewise) momentum spectrum of these elementary quons.

HEISENBERG'S UNCERTAINTY PRINCIPLE

Quantum theory's representation of attributes as waveforms gives us a new way to look at the Heisenberg uncertainty principle (HUP). According to HUP, every attribute A possesses a conjugate attribute V. For any quon, try as you will, you cannot reduce the mutual measurement error associated with these two attributes below a certain natural limit.

We recall from Chapter 5 that for every waveform family W there exists a conjugate waveform family M whose members resemble W waveforms the least. If you use the synthesizer theorem to break any wave w into W waveforms you get a certain spectral width ΔW; if you break this same wave w into M waveforms you get a different spectral width ΔM. The spectral area code, valid for any waveform, requires that:

$$\Delta W \, \Delta M > 1$$

The Heisenberg uncertainty principle follows directly from this spectral area code because the quantum theory's conjugate attributes correspond exactly to conjugate waveform families via the quantum waveform-attribute connection. Thus a quon beam for which a measurement of attribute A gives a narrow attribute spectrum (ΔA small), must give a broad attribute spectrum for a measurement of its conjugate attribute V (ΔV large). The Heisenberg uncertainty principle follows solely from the waveform-attribute connection and has nothing to do with the "unavoidable disturbance of the system by measurement." By virtue of their wave nature, quons simply possess an innate "uncertainty" which makes dual precision measurements physically impossible.

Both Heisenberg and Bohr warned against interpreting the HUP in terms of a measurement disturbance. Rather they claimed that this relation marked the limits beyond which classical notions concerning attributes could not be pushed. One could speak classically about position and momentum only as long as those attributes were not too sharply defined. However, when you imagine conjugate attributes defined with an accuracy greater than that permitted by the HUP, you are thinking about something that cannot exist in nature, like a square circle.

From the point of view of humans attempting to discover as much as possible about nature, the necessary broadness of the mutual spectral widths of conjugate attributes appears to be a limitation on measurement. Hence humans refer to this phenomenon as an "uncertainty principle." However, from the quon's point of view the spectral area code represents a guarantee from nature itself that its "realm of possibilities" will never be diminished past a certain point. If you decrease the realm of a quon's position possibilities, you automatically extend the realm of its momentum possibilities. This natural feature of waves—their inability to be spectrally compressed in two conjugate dimensions—is a boon to quons, ensuring each one its own forever irreducible realm of possibilities. I call the HUP, seen from the quon's vantage, the "law of the realm": Thou shalt not decrease a quon's total realm of possibilities below a certain limit. The law of the realm is no arbitrary decree but a mathematically certain consequence of wave nature itself.

Seen from outside—the human point of view—these obligatory conjugal relations look like "uncertainties." From the inside—the quon's point of view—they feel like "realms of possibility," the basic inalienable estate of every quon in the universe.

HEISENBERG'S PRINCIPLE FOR TWO-VALUED ATTRIBUTES

The Heisenberg uncertainty principle (aka law of the realm) governs all dynamic attributes, even quantized attributes which possess digitally perfect accuracy for a single measurement. It is instructive to see how these two equally fundamental quantum effects coexist—the perfect accuracy of quantized attributes and the obligatory mutual uncertainty of conjugate attributes.

Of particular importance for our purposes are quantized attributes which possess only two values. Two-valued attributes are the simplest possible dynamic attributes, yet they are complicated enough to display a full range of quantum effects.

The HUP for two-valued attributes takes a particularly simple form. If we determine attribute A precisely (it's $A1$ for sure, not $A2$), then its conjugate attribute V must become maximally uncertain. How uncertain can a two-valued attribute get? For an attribute with only two values, "maximally uncertain" means a 50-50 mixture of $V1$ and $V2$, which occur at random, like flipping a fair coin. If we choose to measure attribute V exactly, then its conjugate attribute A displays a 50-50 random mixture. This behavior holds for all two-valued attributes in quantum theory.

Because attributes A and V are quantized, each individual measurement is perfectly exact (either $A1$ or $A2$, for instance). The concept of "uncertainty" applies not to a single measurement but to the measurement of an ensemble of quons. Along with these perfect individual measurements, the HUP requires that neither nature nor an ingenious experimenter will ever succeed in producing a quon beam which always yields $A1$ when the A attribute is measured $(\Delta A = 0)$ and always yields $V1$ when the V attribute is measured $(\Delta V = 0)$.

Quantum theory represents each quon by a proxy wave, and its mechanical attributes by specific waveforms. The proportion of an attribute's waveform which the wave function displays when analyzed with respect to that waveform-attribute family represents the probability of measuring an attribute with that value. This theory deals with the world in a particularly indirect manner. It focuses strictly on measurement acts, not on how the world might behave between measurements; it does not describe single measurement events but only patterns of events, for which it gives merely statistical predictions.

Reality researchers want more. They want to know what the single events which make up the world are doing both between and during measurements. Quantum theory predicts the facts perfectly, but it leaves us in the dark concerning this other kind of reality. Einstein expressed this desire to look behind the facts when he said, "I still believe in the possibility of a model of reality, that is to say, of a theory, which represents things themselves and not merely the probability of their occurrence."

In the next chapter we look at some attempts to go beyond quantum theory to the things themselves.

7

Describing
The Indescribable:
The Quantum Interpretation
Question

They are not smooth-surfaced, rectangular or carbon-ringed units
which fit together like bricks. Each molecule is a heavenly octopus
with a million floating jeweled tentacles hungry to merge.

Timothy Leary

Quantum facts such as the one-hole diffraction experiment suggest that
each electron acts like a wave between observations, but behaves like a
particle whenever it is observed. Quantum theory—the math that describes these facts—likewise reflects each quon's double identity by representing an unmeasured entity by a particular waveshape—whose form
encodes the probability of observing a particle-like event with definite
location and attributes. Both quantum theory and quantum fact support
the notion that the real situation of an unmeasured electron (or any other
quon) is radically different from the reality status of any electron anyone
has ever measured.

Some physicists would like to blame the quantum dilemma on the
observer's inevitable disturbance of what he measures. However, if we take

quantum theory seriously as a picture of what's really going on, each measurement does more than disturb: it profoundly reshapes the very fabric of reality.

We might imagine, as a model of a disturbing measurement, trying to locate accurately a swiftly moving wasp that wants to escape our caliper's jaws: the quantum world—delicate and dynamic—is just too skittish for our clumsy instruments to pin down.

Simple disturbance models of this kind, however, fail to do justice to the quantum description. Quantum theory suggests that before we measure the particle (wasp), it's not a particle at all but something as big as a whale (wave). Calling quon measured a wasp, unmeasured quon a whale still misses the flavor of a quantum measurement because wasp and whale are still both animals. In the quantum description an unmeasured quon does not enjoy the same style of existence as a measured quon; an unmeasured quon dwells in a more attenuated state of reality than the quons which appear on our phosphor screens.

Imagine that the whale dwells not in the real world but on the spirit plane; the wasp is real. Now we are closer to the sense of the quantum description. Whenever a measurement occurs anywhere in the world, something like a ghostly whale (immense, insubstantial, permeable, and wavelike) turns into something like a real wasp (minute, substantial, and particle-like). A quantum measurement resembles good stage magic more than a clumsy meter reading.

In terms of the whale/wasp analogy, the quantum reality question divides into two parts: 1. what is the nature of the whale? 2. how does the whale change into a wasp? The quest to describe the whale is called the "quantum interpretation question." This mystery beast—a quon in its wild, unmeasured state—is represented by a wave function. The interpretation question asks: what does the quon's proxy wave tell us about the factual situation of an unmeasured quon?

The matter of how whale becomes wasp is called the quantum measurement problem: what does the quantum representation of a measurement—as a soft prism splitting a proxy wave into waveforms—tell us about what actually goes on in the measurement act?

I discuss the quantum measurement problem in the next chapter, and consider here the quantum interpretation question. What is the nature of quon unmeasured? Can we describe the world's unseen whales in words or must we remain silent? In between observations a quon is represented by its proxy wave: the wave function is the best clue we have concerning the real nature of the unmeasured universe.

When dealing with a new quantum entity, a theorist's main problem is how to construct its wave function. She commonly resorts to solving some sort of wave equation, such as Schrödinger's equation for slow quons, or Dirac's equation for quons moving near the speed of light.

FEYNMAN'S SUM-OVER-HISTORIES VERSION OF QUANTUM THEORY

In 1948 Richard Feynman, then at Cornell, devised a way of computing a quon's wave function so unusual that it ranks as a quantum-theoretical fourth way, comparable to Heisenberg's matrix mechanics, Schrödinger's wave mechanics, and Dirac's transformation theory. Feynman's method, called the sum-over-histories approach, is useful not only for computations but for the insight it gives into what the wave function might mean.

Feynman was inspired by the work of Christian Huygens, a seventeenth-century Dutch physicist who invented a new way of analyzing light waves by breaking them up into simple sums of spherical wavelets. In the twentieth century Richard Feynman adapted Huygens' technique to quantum waves by breaking them up into simple sums of "elementary histories." As an illustration of his method, let's look at how Feynman might go about constructing the quantum wave for the one-hole diffraction (Airy pattern) experiment. Fig. 7.1 shows the familiar quon gun shooting electrons at a phosphor screen through a circular hole. To calculate the electron's proxy wave ψ, Feynman postulates that the wave amplitude on the screen is equal to the sum of the amplitudes of all possible ways that an electron can get there from the quon gun. Furthermore all paths are equally important, none better than any other. Feynman implements this quantum democracy of possibilities in his scheme by assigning the same amplitude to every path. Each path differs from its fellows only by its phase. A path's phase at any location depends on its history, the route that brought it there.

Feynman assumes that the unmeasured world works according to two rules:

1. A single quon takes all possible paths.
2. No path is better than any other.

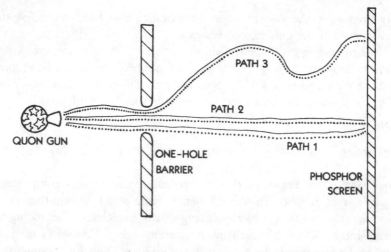

Fig. 7.1   *Feynman's sum-over-histories approach to quantum theory. In Feynman's scheme, a quon acts as though it takes all paths at once. These paths, unlike classical trajectories, possess phases which add wavewise to produce the system's proxy wave—a representation of the probability pattern of a large number of quons prepared in the same state.*

These rules do not mean that an unmeasured electron is free to take any path it pleases. The electron has no choice: if it follows the Feynman rules, the electron is bound to take all paths at once. Furthermore, the quantum democracy of possibilities forbids a quon from "treading harder" on one path and skipping lightly over others. Even though we usually observe an electron near its classical path, that path is no more special in Feynman's scheme than a "wild" path that zigzags crazily toward the screen.

Feynman's idea of getting probabilities by adding up possible paths has much in common with classical statistical reasoning. Gamblers, for instance, compute dice odds by counting how many "dice paths" add up to seven, and so on—the more dice paths there are, the more probable the outcome.

Electrons are odder than dice because: 1. a single electron takes all paths—a single die takes one; 2. electrons have phases and dice don't. If dice (unobserved) took all paths and were equipped with phases, one could imagine a situation where you could throw a six with each die by itself but could never throw boxcars (double sixes) because sixes have opposite phases and cancel each other out.

But if the electron sprawls out across space, going everywhere at once, why does it seem in your TV tube to travel in a straight line? Fig. 7.1 shows three Feynman paths for an electron, two more or less straight paths near the classical trajectory, and one "wild" path.

In the process of adding up his paths, Feynman discovered that next to every wild path runs a parallel path with exactly opposite phase. Since two waves with equal amplitude and opposite phase totally cancel, complete destructive interference removes all wild paths. Only in the vicinity of the classical electron trajectory—in this case the straight line connecting gun, hole, and screen's center—do any paths survive. Because paths add in phase near the classical path, the wave function's amplitude is largest there. Ironically the fact that an electron has phase—a typically quantum property—is what saves the day for classical physics: if electron paths couldn't interfere, the electron would be zipping all over the place.

Feynman's method works: summing up all paths gives the same wave function as solving Schrödinger's equation. For the one-hole diffraction experiment, the sum-over-histories approach yields the familiar Airy pattern centered around the classical straight-line electron trajectory.

Feynman's discovery that he could build a quon's wave function by summing over its possible paths gives us a better view of the wave function's nature than does merely solving an equation. The sum-over-histories approach suggests that the quantum proxy wave represents the totality of possibilities—plus mutual phases—open to a quantum entity. My use of the term "possibility wave" for a quon's wave function and "realm of possibility" for a quon's "uncertainty" is motivated by Richard Feynman's picturesque approach to quantum theory.

THE ORTHODOX ONTOLOGY

Speculation about what actually goes on in the world between measurements goes against the prevailing fashion in physics. Most physicists use quantum theory as a tool for calculating results and "leave it to the philosophers" to wonder what's really going on behind the mathematics. Although physicists officially plead ignorance concerning the ontological status of unmeasured quons, they in fact adhere to an unofficial party line which I call the orthodox ontology. Most physicists accept this ontology without question; only a few mavericks do not.

The orthodox ontology rests on one simple postulate concerning the real physical situation of an unmeasured quon. This postulate cannot be verified experimentally, nor can it be derived logically. It is difficult to say exactly where this postulate comes from, except that it represents most physicists' intuitive feelings about how the quantum world actually works; it summarizes a sense of the very nature of things acquired over years of contact with the details of quantum theory and how it is applied in practice. This postulate may not correspond at all to the reality underneath, but if it doesn't the majority of physicists will have to change their tune.

The central postulate of the orthodox ontology is this: All quons represented by the same proxy wave are physically identical. Two quons represented by the same proxy wave are said to be "in the same state." In terms of quantum states, this postulate reads: "All quons in the same state are exactly alike."

For example, every electron that goes through the hole in the Airy experiment is described by the same proxy wave: all Airy electrons are in the same quantum state. According to the orthodox ontology, before they're measured the physical situation of each of these electrons is absolutely identical: there is no difference whatsoever between electron #123 and electron #137.

Physicists use the wave function to calculate the probability that certain attribute values will be realized upon measurement. Because it deals in probability, the wave function—like dice odds—has an obvious statistical meaning which is relevant to the behavior of a large number of quons. But because of the orthodox identity postulate, the wave function describes a single quon as well. If all quons are physically identical, the distinction between a statistical description and an individual description vanishes.

In the Airy experiment, all electrons do not strike the same phosphor molecule but hit different spots on the screen. If all electrons are really alike, why do they behave differently? The orthodox ontology explains the fact that unmeasured electrons are identical *in being* but different *in behavior* by appealing to quantum randomness. The essence of quantum randomness is simply this: identical physical situations give rise to different outcomes.

Once you get down to the quantum randomness level, no further explanation is possible. You can't go any deeper because physics stops here. Albert Einstein, no fan of the orthodox ontology, objected to this fundamental lawlessness at the heart of nature when he said that he could not believe that God would play dice with the universe. This new kind of ultimate indeterminism may be called *quantum ignorance:* we don't know why an electron strikes a particular phosphor because there's nothing there to know about. When the dice fall from the cup, on the other hand, their unpredictable outcome is caused by *classical ignorance*—by our unavoidably partial knowledge of their real situation.

Because they believe in the orthodox ontology, Bohr and Heisenberg can claim that despite its statistical character, quantum theory gives a complete account of the facts. Critics who object that this theory does not explain the observably different outcomes of electrons in the same state fail to appreciate the nature of quantum randomness: identical situations give different results. That's all there is to it. If the orthodox ontology is a true vision of things, there's absolutely nothing in the unmeasured electron's physical situation that tells where it's going to strike the screen. To demand that quantum theory give such information or be judged "incomplete" is to ask for the impossible. Quantum theory gives the most complete description of the electron's state of affairs consistent with the electron's real nature. To add more would be, as the Chinese say, "to put legs on the snake."

Thus when we see a flash in the Airy experiment we should not imagine that just before the event, a tiny electron was heading for one particular phosphor molecule. According to the orthodox ontology, before a measurement occurs, all of an unmeasured electron's possibilities are live possibilities: just before it strikes the screen, the electron is not headed in a particular direction. If we must talk about it at all, just before it hits, the electron is headed everywhere at once. The rule of the road for unmeasured electrons is this: a single quon takes all paths.

Summing up the orthodox ontology:

1. All quons in the same quantum state are physically identical.
2. The wave function gives a complete account of the physical situation of a single quon.
3. The relationship of the experimenter to an unmeasured quon is one of quantum ignorance: the knowledge he lacks is simply not there to be known.
4. A single unmeasured quon takes all paths open to it.
5. Measured differences between identical unmeasured quons arise from quantum randomness.

THE ORTHODOX ONTOLOGY'S OPPOSITION

Although most physicists accept the orthodox ontology, a small but prestigious minority believe that the world works along different lines. Distinguished dissidents from the quantum identity doctrine include Albert Einstein, Louis de Broglie, Erwin Schrödinger, David Bohm, and John Stewart Bell, as well as many lesser lights. These physicists offer various alternatives to orthodox reality, but most of their proposals involve explaining the world in terms of familiar classical concepts, a quantum reality I've termed neorealism.

Opponents of the orthodox ontology deny its primary postulate and end up with a diametrically opposite view of unmeasured reality which looks like this:

1. Quons in the same state are physically different.
2. The wave function gives a statistical description of an ensemble of quons, and a necessarily incomplete description of a single quon.
3. The experimenter's relationship to an unmeasured quon is one of classical ignorance: certain variables which quantum theory omits are hidden from view.
4. A single unmeasured quon takes one (usually unpredictable) path.
5. Quons in the same state show measurable differences because they were physically different before measurement.

Because of their belief that quantum randomness stems not from utter lawlessness but from hidden causes, these heretics from the orthodox view are sometimes called "hidden-variable" physicists. Their goal is to "complete" quantum theory by constructing a deeper theory which includes an explanation not only of its randomness but of what actually goes on in a measurement.

In a typical hidden-variable model of reality, the world is described in the same manner whether observed or not. The electron in particular is always a particle, just as it seems to be whenever we look at it. However, the electron's motion is controlled by an invisible force field—the so-called pilot wave—whose properties are adjusted to reproduce the experimental facts. In this model of reality, the world is made of both particles and waves, not a single substance that shows both aspects.

Hidden-variable models were originally invented to "complete" quantum theory by accounting for its otherwise inexplicable random results. As a philosophical bonus, such models also describe reality without sanctifying the measurement act, restoring a refreshing sense of objectivity to physics. In the hidden-variable version of reality, measurements are ordinary interactions no different from nature's other interactions. Hidden-variable realities are also entirely picturable in classical terms: there are, for instance, no ghostly white whales swimming around between measurements. Hidden-variable models are philosophically attractive, but they possess one serious drawback which diminishes their value in the eyes of most physicists.

In the orthodox view the world is represented by a fictitious proxy wave with no pretentions to being real. On the other hand, in the hidden-variable models, the wave that tells the electron how to move is considered just as real as the water in the Pacific Ocean or the field surrounding a bar magnet. In order to match the quantum facts, this real wave must possess some quite remarkable properties: primarily it must connect with every particle in the universe, be entirely invisible, and travel faster than light.

Gravity likewise connects every particle in the universe. Physicists are not particularly bothered by this aspect of the pilot wave. The fact that the pilot wave is in principle unobservable (you can only infer its presence from its effect on the electron) disturbs physicists a bit because of the high value they place on observability. However, the fact that the quark may also be unobservable in principle (in most models of elementary particles a quark combines with other quarks too strongly ever to be isolated) has not diminished physicists' enthusiasm for these hypothetical entities. A more serious objection than invisibility is that in these hidden-variable models, an electron's pilot wave routinely travels faster than light.

Most physicists simply refuse to imagine that each electron in the world is guided by an invisible superluminal wave. Bell's theorem shows that all efforts to eliminate the superluminal character of these waves must fail. Bell proves (among other things) that it is impossible to construct a hidden-variable model which explains the facts without including something that goes faster than light.

We acknowledge for the moment the neorealist point of view but will continue to explore visions of reality consistent with the orthodox ontology.

WHAT SORT OF REALITY LIES BEHIND THE WAVE FUNCTION?

The quantum interpretation question is concerned with the nature of unmeasured reality: what is the relationship between reality and its representation, the fictitious proxy wave ψ?

Let's look at some proxy waves for simple quantum systems and ask what's really going on in the world they represent. I assume, in line with the orthodox ontology, that in addition to its statistical meaning the proxy wave gives a complete account of a single (unmeasured) quon. When we look at such a wave function we are seeing a picture (more accurately a representation) of one electron, not a crowd of them. Let's look at some wave functions representing one of nature's simplest systems, the hydrogen atom consisting of a single electron caught in a proton's electron field. Fig. 7.2 shows the wave functions for hydrogen in its ground state and in three of its excited states. Like a phosphor, hydrogen can be excited from its ground state (H) to various high-energy excited states (H_1^*, H_2^*, H_3^*). These excited states last for a few billionths of a second, a comparatively long time on the atomic scale. (Light travels about a foot in a billionth of a second, a distance bigger than a billion atoms.) At a certain unpredictable moment, a particular excited state emits a photon of light and instantly returns to its ground state. According to the orthodox ontology, the fact that different electrons in the same excited state have different lifetimes is not due to hidden differences but to quantum randomness. All hydrogen atoms everywhere in the universe are exactly alike even though each behaves differently.

The sameness of all atoms with a common wave function influences the way our world works. Here MIT physicist Victor Weisskopf points out an important biological consequence of atomic identity:

"No two classical systems are really identical. But in quantum theory it makes sense to say that two iron atoms are 'exactly' alike because of the quantized orbits. So an iron atom here and an iron atom in the Soviet Russia are exactly alike. Our hereditary properties are nothing else than the quantum states of parts of DNA. In some way the reoccurrence every spring of a flower of a certain shape is an indirect expression of the identity and uniqueness of quantum orbits."

Let's look first at the hydrogen ground state H whose electronic wave function takes the shape of a sphere (Fig. 7.2). The sphere's diameter

HYDROGEN PROXY WAVES

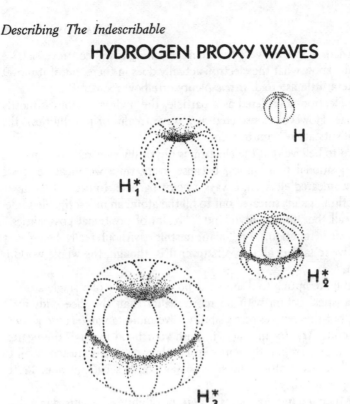

Fɪɢ. 7.2   *Proxy waves for some states of the hydrogen atom: the ground state (H); the first excited state (H₁*); the second excited state (H₂*); the third excited state (H₃*). In addition to standing for the probability pattern of the position attribute of a large number of hydrogen atoms, each of these waveforms represents (according to the orthodox ontology) the most complete description possible of a single hydrogen atom.*

measures the electron's realm of spatial possibilities (also called its "uncertainty"), a quantity previously referred to as ΔX. The extent of the electron's realm of position possibility is just the size of the atom itself, which in this case is about 8 Bohrs (1 Bohr = about 5 billionths of a centimeter). The central nucleus where 99.9 percent of the atom's mass resides is some 200,000 times smaller. If the nucleus were the size of our sun, the hydrogen ground state would be twenty times larger than the solar system!

If the electron were really a point particle moving around in its spatial realm, it would reside in a space so vacant it would make the solar system seem crowded. Since the wave function's meaning is still controversial, we do not really know what the electron actually does in there, but if it moves around like a little asteroid, it has plenty of elbow room.

For the electron considered as a particle, the hydrogen atom is mostly empty space. However, considered as a wave (realm of possibilities), the electron fills its atom brim to brim.

It is hard to believe that the electron is physically smeared out across its realm of positional possibilities, because every time we measure it we never see a smeared electron, always a point particle. In each atom, however, something seems smeared out to fill the atom, an indescribable something we call the "probability cloud," "realm of positional possibilities," "electron wave function," or "quantumstuff" without really being very sure what we're talking about. Whatever it is, though, the whole world is made of it.

It would be tempting to dismiss the possibility wave of a single atom as an airy statistical fiction with no more reality than the dice odds for a single roll, but these waves of possibility have more tangible consequences than dice odds. Try, for instance, to push your hand through the nearest wall. Since atoms are mostly empty space, their electrons are too small to stop you. Only each atom's possibility wave pushes back at you. Pretty substantial, aren't they?

Despite their representation as "mere possibilities," atoms don't drift through one another like mist but stack up more like billiard balls, each one pressing tight against his neighbor's probability cloud. Whatever the actual nature of the ψ wave for an individual system, it is evidently something solid enough to sit on.

The hydrogen ground state wave function H is spherical. Excited state H_1^* looks like a pumpkin; H_2^* like a pair of earmuffs. Excited state H_3^* looks like one doughnut on top of another.

Perhaps we could imagine that the electron is a busy insect that traverses its spatial realm so swiftly that it seems to be everywhere at once. The electron's proxy wave would then be a sort of blurred time-exposure photo of the electron's actual motion. We recognize this supposition as a simple neorealist (hidden-variable) interpretation of the wave function. The hidden variable in this case is the electron's definite but constantly changing position. In this conjectural interpretation, all hydrogens are not the same but differ on account of the momentary position of their electrons.

One obvious difficulty with this hidden-variable model is that some proxy waves—H_3^x, for instance—consist of two disjoint parts. For excited state H_3^x, the probability that the electron will be seen in the upper doughnut is $1/2$, in the lower doughnut also $1/2$. But the probability for the electron to be found anywhere on the plane separating the two doughnuts is exactly zero. This hypothetical insect electron must spend half its time in each doughnut but can never be caught in between.

Some people imagine that the quantum paradoxes result from the disparity in size between the human and quantum worlds: our big world works one way, the Lilliputian world of the atom works by different rules. Russian-American physicist George Gamow, in his amusing book *Mr. Tompkins Explores the Atom*, created a world in which Planck's constant, which sets the scale for quantum phenomena, is so huge that quantum effects are commonplace in everyday life. In Gamow's fabulous world, for example, Mr. Tompkins watches a billiards game in which the cue ball bounces off the target ball and recoils *in all directions at once*. Like Max, our imaginary experimental physicist, Mr. Tompkins experiences firsthand the quantum world's democracy of possibilities: he actually sees a single quon taking all possible paths.

Suppose we could shrink ourselves and our apparatus to nuclear dimensions and enter hydrogen wave function H_3^x like a spaceship exploring an unknown solar system. While hovering inside the upper doughnut of this excited state we deploy our electron detector, which consists of the usual phosphor screen but is made of the same condensed matter—a piece of neutron star perhaps—as our spaceship. (We will have to make our measurements in a hurry; this state has an average lifetime of only six billionths of a second.) Could we solve the mystery of what the wave function really represents if we were small enough to get inside an electron's proxy wave?

What would we see in there? Nothing special, I believe. The hydrogen atom has one electron. If we are lucky it will make a single, well-localized flash on our condensed matter screen. That's all. Electrons are always observed to be particles no matter what the size of the measuring apparatus. The quantum paradoxes do not arise because of the relative sizes of atoms and humans (a mere quantitative difference), but because of a more fundamental qualitative difference between the experiences of human and atomic beings. As Bohr and Heisenberg have pointed out, human experience is inevitably classical, but atoms do not exist in a classical manner.

Indeed, we already know what it would be like to live inside an atomic wave function, because we walk around "inside atoms" whenever we go outdoors on a starry night.

STARLIGHT PROXY WAVES

The size of an atom is equal to its realm of spatial possibilities—no more than a few billionths of an inch. Excited atoms may puff up to more than a thousand times this size, still too small to see. On the other hand, the spatial realm of a single photon from a distant star can vary from a few feet in diameter to an area the size of Texas.

The spatial realm of each star's photons depends inversely on that star's apparent angular size. Small faraway stars have an enormous realm; the realms of big nearby stars are smaller. The smallest realm in the sky belongs to Betelgeuse, a giant red star in Orion's shoulder, whose realm on Earth is approximately ten feet in diameter. This means that the proxy wave of each photon from Betelgeuse is bigger than a bathtub.

The Heisenberg uncertainty principle explains the large spatial realms of starlight. Imagine a beam of photons leaving Betelgeuse (520 light-years away) heading in Earth's direction. Unless these photons possess extremely small sidewise motion, they will drift out of the beam during their 500-year transit time. To travel from a distant star and strike a target as small as your open eye, a beam of starlight must possess a very tiny spread in sidewise momentum. According to Heisenberg's principle, a small spread in (sidewise) momentum requires a compensatory large spread in (sidewise) position. That's why the photons from Betelgeuse are ten feet wide. Other stars have wider wave functions.

How thick are these photon proxy waves from the stars? Applying the uncertainty principle in the photon's direction of motion, we calculate a realm of positional possibility thinner than a soap bubble. These stellar proxy waves have the look of very wide, bubble-thin pancakes hurtling through the night sky at the speed of light.

Living inside a photon's proxy wave we notice nothing unusual. In particular, a photon detector (your eye, for instance) exposed to light from Betelgeuse does not see a thin ten-foot-wide luminous disk. All that's ever seen by eye or phosphor screen is an occasional flash of light. Every direct measurement of light reveals it to be a point particle—the measured pho-

David Finkelstein, head of the School of Physics at Georgia Tech, is mapping the geography of a quantum-logical world—a world built from "quons" possessing definite attributes which combine according to non-Boolean rules (QR #5).

*John Wheeler,
an active proponent of observer-created reality (QR #2).
Under his guidance the Institute of Theoretical Physics at the University of Texas has grown into an important center for quantum reality research.*

Hugh Everett III,
after originating the many-worlds
interpretation (QR #4),
joined the Pentagon's Institute for
Defense Analysis where he developed the
widely used "Everett algorithms"
for optimizing strategic decisions.

John Clauser,
now researching fusion power at Lawrence
Livermore Laboratory,
was the first to verify
Bell's theorem experimentally thereby
factually establishing non-locality as
a general feature of nature.

David Bohm,
of London's Birkbeck College,
has explored many facets of quantum
reality during his long career,
including the Copenhagen interpretation
(QR #1 and #2),
neorealism (QR #6),
and quantum wholeness (QR #3).

Henry Pierce Stapp,
a theoretical physicist at Lawrence
Berkeley Laboratory,
propounds a consciousness-created reality
(QR #7) based on American philosopher
Alfred North Whitehead's world view—
a universe-in-process full of surprises,
whose history unpredictably unfolds
as the result of the activities
of myriads of conscious beings.

ABOVE, *Niels Bohr,*
founder of the Copenhagen interpretation
(QR #1 and #2) which holds that
despite its manifest statistical character,
quantum theory is a complete
theory of nature. If it's not in this theory,
says Bohr, it's not in this world.
Bohr's institute, housed in buildings
donated by Carlsberg Brewery,
became, during the twenties and thirties,
the world center for development
of the New Physics.
(AIP Niels Bohr Library photo)

LEFT, *Werner Heisenberg discovered the*
famous uncertainty principle, which
limits our knowledge of the simultaneous
values of conjugate attributes.
As a way of thinking about quantum reality,
Heisenberg proposed (QR #8)
that a quantum entity's unobserved
attributes are not fully real but exist
in an attenuated state of being called
potentia until the act of observation
promotes some lucky attribute
to full reality status.
(AIP Niels Bohr Library photo)

ton, like the measured electron, always seems to have zero radius. If we could crawl inside a hydrogen atom, I'm sure we'd see the same thing. Individual electrons, like individual photons, always appear as particles to our instruments, even in cases where their wave functions are bigger than the apparatus which measures them.

Despite the fact that our detectors never report anything but pointlike particles, physicists have figured out how to measure the size of these stellar proxy waves.

Deep in the Australian Outback, three hundred miles northwest of Sydney in the village of Narrabri, British physicist Robert Hanbury Brown built the world's first stellar intensity interferometer. He completed this instrument in 1965 and used it to determine the angular diameter of stars by measuring the widths of their photon proxy waves.

The apparent angular diameter of a star measures how much space it takes up in the sky. The sun and moon, for instance, each take up about half a degree of skyspace. (The width of the sun's photon proxy wave is a small fraction of a millimeter.) The angular diameter of stars is too small to measure directly; all stars appear as points even in the largest telescopes. Because of their small apparent sizes, the diameters of stars can be measured only indirectly, by instruments such as the stellar interferometer.

Hanbury Brown's interferometer consisted of two twenty-foot searchlight mirrors mounted on railroad tracks. Each searchlight collected light from the same star and focused it onto a photon detector. At Narrabri signals from each photon detector were combined electronically, but the operation of this instrument is easier to visualize if we imagine that each mirror directs its beam of starlight to a giant phosphor screen (Fig. 7.3). When the searchlight mirrors are close together, an interference pattern appears on the phosphor screen, as light from one mirror shifts in and out of phase with light from the other mirror. Still focused on the same star, the mirrors are rolled apart. The interference pattern disappears when the mirror separation is so large that the mirrors lie outside the photon's wave function. Some stars' wave functions are bigger than Hanbury Brown's railroad line, which was capable of measuring wave functions as large as four hundred feet in diameter—the size of the photon proxy wave from Zeta Orionis, the brightest star in Orion's belt. Stars whose angular size is smaller than Zeta Orionis possess correspondingly larger proxy waves. The wave functions of some very distant stars are considerably wider than the Australian continent.

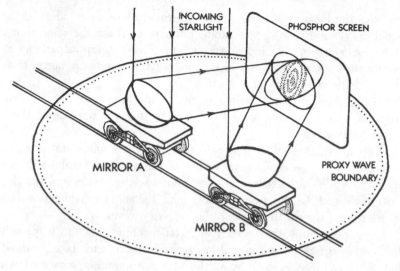

FIG. 7.3   *The stellar interferometer, measuring the size of a photon's proxy wave. Each searchlight mirror focuses light from the same star onto the phosphor screen. An interference pattern is observed as long as the two mirrors lie inside the photon's proxy wave, indicated by the solid/dotted circle. When the mirrors move outside this circle—the boundary of the photon's wave function—the interference pattern disappears.*

On the phosphor screen, the interference pattern is formed one photon at a time: photons are undoubtedly point particles. Yet the pattern itself can only be explained by something wavelike—represented by the photon's proxy wave—which is able to bounce off both mirrors at once and come together on the screen to produce the zebra-striped pattern characteristic of wave interference. Like the one-hole diffraction experiment, the stellar interferometer demonstrates the coexistence of wave and particle effects. Nothing but particles are ever detected directly, but the pattern of these particles must have been caused by some sort of wave—the form light seems to take when it is not being measured. The interferometer actually determines the boundaries of this quantum wave and shows that the size of such waves—the same kind of wave an atom is made of—need not be restricted to atomic dimensions.

To understand quantum theory, as opposed to merely knowing how to use it, we must answer two questions: 1. What does the wave function really mean? (the interpretation question); 2. What happens during a quantum measurement? (the measurement problem). The quantum reality crisis arises because physicists have no good answers for either of these questions. In this chapter we looked briefly at the interpretation question. We turn next to the measurement problem.

8

"And Then A Miracle Occurs": The Quantum Measurement Problem

No language which lends itself to visualizability can describe the quantum jumps.

Max Born

A gravity wave is a ripple in the curvature of space-time. Einstein's general theory of relativity predicts that gravity waves ought to be generated wherever huge masses accelerate—for instance, in binary star systems. If Einstein is right, gravity waves from all parts of the sky pass through the Earth every day.

A gravity wave slightly warps every object in its path, squeezing it in one direction and stretching it in the orthogonal direction. Because of the ubiquity of gravity waves, every object we see is continually pulsing to the gravitational rhythm of distant stars. However, compared to the electric force that holds things together, gravity is exceedingly weak; consequently the daily deformation of ordinary objects by gravity waves has escaped detection.

SOVIET GRAVITY WAVE DETECTOR

At Moscow University, Vladimir Braginski is looking for gravity waves by monitoring tiny changes in the shape of a 200-pound sapphire crystal cylinder. Braginski chose this exotic material because after being struck it continues to quiver for a record time. Sapphire's long ringing time permits making a maximum number of measurements before the gravity wave's impact fades away. To isolate it from terrestrial noise, the Soviet sapphire is suspended by wires in a vacuum chamber and cooled to near absolute zero.

So far, Braginski and his colleagues haven't seen any gravity waves and are working to decrease the noise in the position detectors which sense the distortion of the sapphire rod. If they can reduce this noise by several orders of magnitude, the Russian physicists expect to reach a point where the sensitivity of their measurements is limited by the quantum nature of the sapphire bar.

The sapphire crystal, considered as a quantum entity, is subject to the Heisenberg uncertainty principle: any measurement which accurately determines its position attribute must widen its momentum's realm of possibility. In general the uncertainty principle does not limit how accurately you can measure *a single attribute* but only restricts the mutual accuracy of measurement of *two conjugate attributes*. As far as the uncertainty principle is concerned, you can make a position measurement as accurately as you please. Since Braginski's group is only interested in the position of the sapphire bar—more precisely just the position of the boundaries that define its shape—one would not expect them to be troubled by a variance in the bar's momentum.

The Russians could indeed ignore the bar's momentum spread if they planned to make only one position measurement. However, for reliability's sake they must measure position repeatedly; the more measurements the better. Best is when the bar's position can be monitored continually.

The first accurate position measurement induces via the uncertainty principle a large momentum spread. For the same reason a collection of particles with different momenta will quickly drift apart, this induced spread in the bar's momentum soon results in a spread in the bar's position. Momentum just happens to be an attribute whose uncertainty feeds back into the position attribute. Braginski calls such a situation—where accurate measurement of one attribute is spoiled by the back-reaction of the Heisenberg spread in its conjugate attribute—*a quantum demolition measurement.* For such measurements the quantum nature of the sapphire bar sets an ultimate limit to the accuracy with which you can know the bar's position.

Braginski and other theorists, however, see a loophole in quantum theory which will allow them to get around this fundamental limit. Besides position and momentum, the sapphire bar possesses an infinity of external attributes—one for every waveform family in its proxy wave's configuration space. Though each attribute and its corresponding conjugate attribute are mutually constrained by the uncertainty principle, *not every attribute is of the demolition type.*

Braginski and his colleagues are particularly interested in an attribute called X_1, which may be thought of as consisting of half position, half momentum combined wavewise with a certain phase relationship. Attribute X_1's conjugate attribute, called X_2, is also composed of half position, half momentum, but added together with a different phase. Like all conjugate attributes, the realms of possibilities of X_1 and X_2 are jointly constrained by the uncertainty principle, but unlike position and momentum, the spread of possibilities of one of these attributes confines itself to that attribute and does not feed back into its conjugate attribute.

Since Heisenberg uncertainty in attribute X_2 does not contaminate its conjugate attribute, the quantum nature of the sapphire bar places no limit on repeated measurements of X_1. Accuracy is limited now only by the experimenter's ingenuity. Braginski calls such a potentially error-free observation a quantum non-demolition measurement (QND measurement). In addition to improving future gravity-wave detectors, physicists plan to use QND measurements to reduce Heisenberg uncertainty noise in laser beam communicators.

Because X_1's realm of possibility can be made as narrow as you please and kept that way, the particular waveforms which correspond to QND attributes such as X_1 and X_2 have come to be called "squeezed states."

Braginski's proposal to detect gravity waves in Moscow by measuring a non-demolition attribute (squeezed state) rather than the more conventional position attribute illustrates two important aspects of a quantum measurement: 1. Quantum theory applies in principle to all physical entities no matter how large; 2. A crucial step in any quantum measurement is choosing which attribute you will look at.

Although originally devised to deal with tiny invisible atoms, quantum theory doesn't quit when things get big and see-able. We've already met up with starlight proxy waves bigger than parking lots. Now Russian physicists reckon a king-sized sapphire as much a quantum entity as a photon of light.

THE QUANTUM METER OPTION

I call the experimenter's ability to freely select which attribute he will measure the quantum meter option. In the case of the Moscow gravity wave detector, choosing to measure attribute X_1 rather than position leads to a more accurate knowledge of the sapphire bar's deformation. Exercising your meter option is a necessary part of any quantum measurement.

There are two senses in which an observer may be said to "create reality" when he makes a quantum measurement: the first kind of reality creation occurs whenever an observer exercises his meter option; observer-created reality of the second kind takes place when the observer "collapses the wave function." Most claims for an observer-created reality concern acts of the first kind.

Because quantum theory represents a quantum entity by a wave and attributes as waveforms, it's easy to see how the observer may be accused of "creating reality" whenever he chooses which attribute he will observe. The key operation in quantum measurement is selecting a waveform family prism with which to analyze the system's wave function. Which prism you select determines what attribute you want to look at.

Imagine doing such a waveform analysis on an ordinary wave—the traffic noise from a busy street corner, for instance. Exercising our meter option, we will analyze this noise in terms of *tuba* waveforms. If we want to look at our sound sample this way, we will observe that traffic noise consists of certain percentages of tuba notes of various frequencies connected by particular phase relations. But is traffic noise *really* an orchestra of tubas? Of course not.

Traffic noise supplies the raw material for this measurement, but the choice of component waveforms is up to the observer. Finding momentum in an electron wave is like finding tubas in a traffic wave. We measure a certain electron momentum only because (thanks to the meter option) we helped put momentum there. Electrons cannot really be said to have dynamic attributes of their own. What attributes they seem to have depends on how we choose to analyze them. A clock comes apart in only one way: it's made of definite parts. A wave, on the other hand, doesn't have parts; you can divide it up any way you please. However, none of these divisions is there to begin with; the kind of parts a wave seems to have depends on how *we* cut it up. The world's wave nature makes us in a certain sense co-creators of its attributes.

There is, however, one sense in which a quantum system may be said to possess attributes of its own. This is the same sense in which traffic noise possesses its own identity no matter how we choose to analyze it. Traffic noise is not made by tubas, it is made from traffic. If we analyzed such noise with a traffic-waveform prism, it would not split into components. In Chapter 5, I called such prisms which do not split the waveshapes they analyze "kin prisms."

Suppose we analyze an electron beam with a momentum prism and its proxy wave does not split. This means that its wave function is a pure sine wave (quantum code for momentum). Experimentally we would observe a beam of such quons to possess a single precise value of momentum. In this special case we could say that every electron in the beam has a specific momentum, which the observer does not create. In other words, in the exceptional case of observation with a kin prism, we might say that the experimenter is seeing not what he put in but what is really there. However, although these electrons seem to possess momentum in a manner reminiscent of classical objects, none of their other attributes is single-valued. All the other attributes come about via the quantum meter option —observer-created reality of the first kind.

To measure a quon beam whose dynamic attributes are single-valued is a relatively rare occurrence. Most quantum measurements give some spread in their outcomes. For these cases the observer may be said to partially create the attributes he observes in much the same way that we can find tuba waveforms in traffic noise if we look for them. Using the synthesizer theorem, we can express a quon's situation in terms of an infinity of different attribute waveforms; the choice is up to the observer. Physicists choose certain attributes so often to characterize physical systems that they deserve to be called major attributes. All others I call minor attributes.

In *configuration space*, where the waveforms dwell which represent a quon's *external motion*, the major attributes are position and momentum. Minor attributes in configuration space include the fanciful "piano attribute" and Braginski's QND attributes X_1 and X_2.

In *spin space*, where the waveforms dwell which represent a quon's *internal motion*, the major attributes are the spin orientations S_x, S_y, S_z in three orthogonal directions. For light and many other quantum entities, the most important minor attribute in spin space is polarization.

THE POLARIZATION ATTRIBUTE OF A LIGHT BEAM

As a concrete example of the quantum measurement problem, let's look at how an experimenter might go about measuring a light beam's polarization. Polarization is the simplest type of dynamic attribute because it can take only two possible values. One less value would turn it into a static attribute. However, despite its simplicity the polarization attribute is complex enough to illustrate the full range of quantum perplexities.

Polarization is an attribute connected with a particular direction in space. For each direction a single photon has only two options: either it is entirely polarized in that direction or it is entirely polarized at right angles to that direction. The only polarization directions that concern us here are the orthogonal—those at right angles to the light beam's direction of travel. To visualize these polarization directions, imagine the light beam encircled by a clock dial. The hour hand's directions represent all possible polarization directions. If we let twelve o'clock equal zero degrees, these polarization directions can be described either by a clock time or by an angle or by conventional direction labels such as horizontal or vertical. Twelve o'clock, for instance, stands for vertical polarization; three o'clock for horizontal polarization. Only half the clock face represents a unique polarization: nine o'clock and three o'clock, for example, both represent the same direction, namely horizontal polarization.

To visualize a polarization measurement, imagine that you are a batter standing on home plate trying to hit photons that the pitcher is throwing. (This pitcher's fast ball is *really* fast: it travels at the speed of light.) You hold your bat at a particular angle and if the photon is polarized at that angle, it's a hit; otherwise it's a miss and you know that the photon was polarized perpendicular to your bat.

A binary outcome connected with a single direction is the most that quantum theory permits you to know about the polarization attribute of a single photon. All you can learn when you make such a measurement is whether the photon is polarized along the direction of your choice (hit), or at right angles to your chosen direction (miss). Your quantum meter option for a polarization measurement consists of choosing for each photon the angle at which you will hold your bat. Because you can only hold the bat in one direction at a time, you cannot find out very much about the overall polarization of a single photon, but you can examine the overall polarization of a *beam* of photons by holding your bat at different angles for each photon in the same beam.

For all light beams considered here the photons will be in the same quantum state, represented by the same proxy wave. According to the orthodox ontology, such photons, before they are measured, are absolutely identical. The orthodox ontology claims that measuring a beam of such photons tells us everything that can possibly be known not only about the beam but about each photon in the beam. Each bat angle ϕ corresponds to a different polarization attribute which I designate by $P(\phi)$. To measure the $P(\phi)$ attribute of a particular light beam, hold the bat at angle ϕ and record the number of hits and misses.

Three kinds of polarized light are particularly notable: 1. If you record 100 percent hits, the beam is said to be completely polarized in the ϕ direction. 2. If you record 100 percent misses, the beam is completely polarized in the orthogonal direction. 3. If you record 50 percent hits/50 percent misses, the light beam is *unpolarized* in the ϕ direction. If the measurement results don't fall into any of these categories, the light beam is partially polarized. Light from the blue sky and light reflected from the sea are instances of partial polarization.

A light beam measured to be unpolarized in every direction is said to be completely unpolarized. Light from the sun, from light bulbs and street-lamps is completely unpolarized.

The uncertainty principle applied to the polarization attribute requires that a light beam can be completely polarized in only one direction: if $P(\phi)$ measures 100 percent hits for a particular angle ϕ, then all other angles must give less than 100 percent hits. Laser light and light that has passed through a Polaroid filter (used in sunglasses and 3-D movie viewers) are examples of light that's completely polarized in one direction.

To illustrate the quantum measurement problem, I consider just two polarization attributes, namely $P(0)$, where the bat lies at zero degrees, and $P(45)$, where the bat is held at 45 degrees. When the bat is held at zero degrees, we measure $P(0)$, which can take only two values: either a hit indicating a vertically polarized photon *(V* photon), or a miss indicating a horizontally polarized photon *(H* photon). When the bat is held at 45 degrees, we measure $P(45)$, which can take two values: either a hit indicating a diagonally polarized photon *(D* photon) or a miss indicating a slant polarized photon *(S* photon). Diagonal and slant are conventional names for the 45-degree and −45-degree directions respectively.

The object of our quantum measurement is a beam of light that is completely *D*-polarized. Every photon in this beam is a *D* photon. Experimentally this means that selecting 45 degrees for your meter option will give you 100 percent hits. We will not make such a trivial measurement but will choose instead to measure this beam's polarization at zero degrees.

Quantum theory tells us that the waveform *D* which represents a completely *D*-polarized beam can be analyzed into a superposition of equal parts *H* waveform and *V* waveform added together with a particular phase. We might represent this waveform relation as:

$$D = H \oplus V$$

where ⊕ stands for wave-wise addition.

This mathematical relationship between polarization waveforms means that if we put waveform *D* into a wave analysis prism whose outputs are *H* and *V* waveforms, the prism will divide the *D* wave into an *H* wave and a *V* wave with equal amplitudes.

The quantum measurement that corresponds to this waveform analysis consists of sending a purely *D*-polarized beam into an attribute sorter which separates any light beam into *H* photons and *V* photons. Such a sorter corresponds to holding the bat at zero degrees, measuring what we have called the *P(0)* attribute.

Because the *D* waveform consists of equal parts *H* and *V* waveforms, quantum theory predicts (via the waveform-attribute correspondence), that a *P(0)* measurement on a pure *D*-polarized beam will give a 50-50 mixture of *H* and *V* photons; the *D* beam will appear to be *unpolarized* in the zero degree direction. Inquiring as to how nature actually contrives to produce this 50-50 mixture lands us quickly and deeply in the very heart of the quantum measurement problem. Before discussing the measurement problem, I describe how an actual polarization measurement might be carried out with the hardware that corresponds to the "bat" in my baseball analogy.

HOW TO BUILD A POLARIZATION METER

An actual polarization measurement closely resembles holding a bat at a particular angle and recording hits and misses. Instead of a bat, physicists hold a crystal of calcite across the light beam.

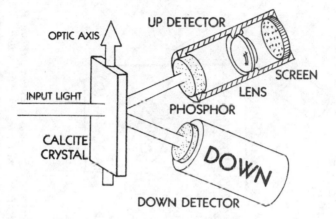

FIG. 8.1 *Construction of a P meter. To measure the polarization attribute P(0) (called variously vertical polarization, zero degree polarization, or twelve o'clock polarization), a calcite crystal whose optical axis points in the vertical direction splits a light beam into two beams that contain photons all polarized along the optic axis (up beam) or all polarized at right angles to the optic axis (down beam). Photon detectors consisting of a phosphor plus a light-sensitive screen record which channel each photon actually takes. The measurement result consists of the number of counts in the up and the down channels.*

In the laboratory the light beam goes through a calcite crystal whose optic axis is pointing in a particular direction. Calcite is a birefringent crystal: it divides light into two beams, the up beam consisting of photons polarized along the optic axis, the down beam consisting of photons polarized at right angles to the optic axis. More about this remarkable mineral may be found in the appendix to this chapter.

To record whether a particular photon actually goes up or down, a simple photon detector is placed in each channel. This detector consists of a phosphor screen that gives off a flash of light when excited by a photon,

A. MEASURING THE P(0) ATTRIBUTE

B. MEASURING THE P(45) ATTRIBUTE

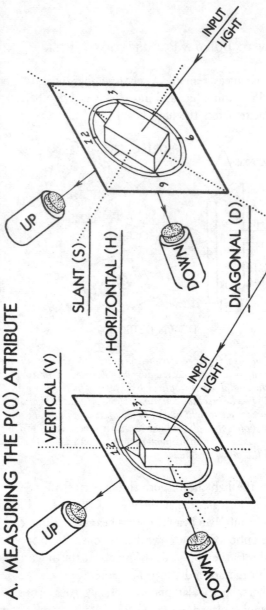

FIG. 8.2 Measuring the polarization attribute P(0) and polarization attribute P(45). When the calcite crystal is set at zero degrees, the P meter is measuring the P(0) attribute. When the crystal is turned to 45°, it measures the P(45) attribute. Polarizations measured in these two experiments define the four photon polarization directions H, V, D, and S discussed in the text.

plus a lens that gathers this phosphor light and focuses it onto a sensitive plate (similar to an element of a solar cell) which produces an electric pulse in response to light. Fig. 8.1 shows a complete polarization meter set to measure the polarization attribute *P(0):* the calcite's optic axis is aligned at zero degrees. The electric pulse from the photon detector is amplified and sent to a recorder (not shown) which prints a *U* or a *D* on a record tape depending on whether the up or down detector fired for that particular photon. Fig. 8.2 shows this polarization meter in two of its many possible settings. When the calcite's axis is set at twelve o'clock (zero degrees), the Polarization meter measures attribute *P(0):* up and down signals from the counters correspond to *V* and *H* photons respectively. When the calcite's axis is set at 2:30 o'clock (45 degrees), the Polarization meter measures attribute *P(45):* up and down signals now correspond to *D* and *S* photons.

SIX VARIATIONS ON THE QUANTUM MEASUREMENT PROBLEM

After this brief digression into experimental physics we are ready to tackle the measurement problem. Start with a pure *D*-polarized beam and measure its *P(0)* attribute with a calcite crystal set at zero degrees. The beam divides into equal numbers of *H* and *V* photons. But what's actually going on here when one beam splits into two?

In terms of this simple experiment—the measurement of *H/V* polariza tion on a pure *D*-polarized beam—it's easy to state the quantum measurement problem. In fact since this problem is so important for the quantum reality question I express it here in six different ways. Although simple to state, this problem is not so easy to solve. All solutions to the measurement problem that physicists have so far come up with either lead to bizarre realities and/or sanctify some aspect of the measurement act.

It's hard to believe that nature endows the act in which humans make contact with quantum entities with a special status not granted to all the other interactions in the universe. Any interpretation of measurement which attributes supernatural powers to the act itself must be regarded with suspicion. There's something philosophically fishy about a measurement-centered cosmos. How the world appears to us must certainly depend on how we measure it, but it's absurd to believe that how the world actually *is* is determined by human observational capacities. "Measurements are happenings," says U.S. Navy physicist T. E. Phipps, "but they are the least of the happenings that go on in the world. It demeans physics, not to mention the world, to shackle physics with its own instrumental tools through any linguistic implication that measurements are *all* that happens in the world."

Here then are six variations on the quantum measurement problem that has baffled physicists for more than half a century.

1. Physicists can't represent a quantum system's physical situation in classical terms; they express it in terms of *quantum possibility waves*. On the other hand, the measuring device and its result can't be expressed in terms of possibilities, but like any other aspect of human experience must be described in terms of *a concrete classical actuality*. The quantum measurement problem is this: at what point between the input photon and the observation of a definite mark *(D,* for instance) on the output tape does the transition occur between these two strikingly different styles of existence? *Where do we put the "cut" which divides the quantum and classical world?*

2. According to the orthodox ontology, because they are all in the same quantum state, the physical situation of each *D* photon is identical to that of every other. However, the results printed on the measurement tape are not all the same. *At what point in the measurement process do identical quantum entities develop differences?*

3. According to the orthodox ontology, after being split by the crystal, each *D* photon dwells in a state of pure possibility. But the measurement results themselves are actual facts, not possibilities. *At what point in the machinery does possibility change into actuality?*

4. According to the orthodox ontology, each *D* photon takes both paths through the analyzing crystal and simultaneously occupies both the *H* and *V* channels. Yet for each photon only one channel actually fires (either *H* or *V* but not both) and prints a character on the output tape. *In the process of quantum measurement, when do two paths turn into one?* And what happens to the path not taken?

5. According to the orthodox ontology, our relationship to a quon's physical situation after it's split by the analyzing crystal is one of quantum ignorance: we do not know the difference between quon #123 and quon #137 because no such difference exists in nature. Not even God can tell apart two quons in the same state. Yet after 1,000 photons have been

measured we are convinced that the measurement tape actually contains marks that are not the same for each photon. In other words, before we look at the tape, our relationship to these results is one of classical ignorance. *How and when does quantum ignorance turn into classical ignorance?*

6. My final variation on the theme of quantum measurement poses the question: *how and when does the wave function "collapse"?* We will learn later about the process of wave function collapse in connection with John von Neumann's all-quantum model of the measurement act.

Notice that the measurement problem seems mostly to arise from taking the orthodox ontology seriously and trying to reconcile its teachings with what is actually observed. One easy solution to the QMP might be to simply *deny the orthodox ontology* and accept a neorealist model of the world. There is no measurement problem in neorealist models, but they have another problem instead. Bell's theorem tells us that no neorealist model will work unless it contains real but invisible faster-than-light force fields, a situation most physicists consider unacceptable.

Those who embrace the orthodox ontology (the majority of physicists) fall into two camps: the followers of Bohr and Heisenberg (Copenhagenists) and the followers of John von Neumann.

THE COPENHAGEN PICTURE OF QUANTUM MEASUREMENT

The Copenhagenists consider ordinary experience the primary unanalyzable reality in terms of which they explain the atomic realm. For Bohr and Heisenberg the world is forever divided into two types of reality: quantum reality which we can never experience, and classical reality which is all that we can ever experience. Quantum theory is not a representation, much less a description, of quantum reality, but a *representation of the relationship* between our familiar reality and the quon's utterly inhuman realm. As Heisenberg puts it: "The Copenhagen Interpretation regards things and processes which are describable in terms of classical concepts, i.e., the actual, as the foundation of any physical interpretation." Harvard physicist Wendall Furry echoes Heisenberg's elevation of ordinary reality to top-dog status: "[In the Copenhagen interpretation] the existence and general nature of macroscopic bodies and systems is assumed at the outset.

These facts are logically prior to the interpretation and are not expected to find an explanation in it."

In other words, the Old Physics attempted to explain macroscopic objects in terms of the atoms which make them up; the New Physics explains atoms in terms of macroscopic objects. In this inverted Copenhagen scheme, there is a sense in which *atoms are made of measuring instruments* and not the other way around. As Heisenberg writes: "Only a reversal of the order of reality as we have customarily accepted it has made possible the linking of chemical and mechanical systems of concepts without contradiction." Of course Copenhagenists believe that their instruments are made of atoms just like everything else, but this manner of thinking (instruments made of atoms) can be carried only so far.

A curious feature of the Copenhagen interpretation is that it considers both the atom and the measuring device to be incomprehensible. We cannot understand the quantum world because its nature is utterly alien to human thought; we cannot explain the classical world because quantum theory—the physicist's only basis for explaining anything today—simply takes the existence of the classical world for granted. In Furry's words, the classical world is "logically prior" to quantum theory and "is not expected to find an explanation in it." Quantum theory predicts how a classical measuring instrument will respond to a quantum system, but the theory itself does not contain such measuring devices—nothing in there but proxy waves. Fortunately for the practice of physics, each of us is born into a world already inhabited by these inexplicable measuring devices: your eye is one example.

According to Bohr, quantum theory describes neither the quantum system nor the measuring device. Quantum theory applies to the *relationship* which exists between these two conceptually opaque kinds of being. Since the measuring device cannot in principle be analyzed, it's the perfect place to put the solution to the measurement problem. In the Copenhagen interpretation, all the mysterious transitions between the quantum and classical kinds of being occur inside the measuring device or more properly at the boundary between measuring device and quantum system. We see that the Copenhagen interpretation does not so much solve the measurement problem as conceal it. It sweeps this problem under the rug, into the one place in the world inaccessible to human scrutiny—the insides of measuring devices.

The Von Neumann Picture of Quantum Measurement

The fact that Copenhagenists must divide the world into a classical part (measuring device) and a quantum part (measured system) displeased John von Neumann. The world obviously has only one nature, and *that nature is not classical*. To reflect the world's necessary unity, von Neumann undertook to represent the whole world one way: he symbolized both the system and the measuring device with proxy waves. Von Neumann's world is entirely quantum—there's not a bit of classical physics in it. Von Neumann described this all-quantum picture of the world in his quantum bible, *Die Grundlagen*. It works. It's possible to represent the entire world (both system and measuring device) in terms of proxy waves if you make one assumption. Von Neumann's crucial assumption is the basis of my sixth (and last) variation on the quantum measurement problem.

As Feynman showed in his sum-over-histories version of quantum theory, one way to think about what unmeasured quons are doing is to imagine that each quon takes all paths. Feynman's picture amounts to a sort of quantum law of motion, analogous to Newton's law of motion for classical objects.

Each quon moves from one state of being to the next according to a ruthless territorial imperative demanding that it occupy all its possibilities at the same time. The fact that most of these paths are obliterated by destructive interference in no way alters a quon's primal orders: fill the Earth with your essence! The law of the realm ensures that no matter how many of a quon's possibilities are destroyed by wave interference or by measurement, a certain minimum remnant will always remain. A quon always possesses, no matter what its circumstances, a realm of possibilities at least equal to Planck's constant of action.

This is the quantum law of motion: Increase and multiply: starting from your inviolable realm, take all possible paths open to you. The natural evolution of a quon's proxy wave is to expand without limit. However, in the measurement act a quon can't realize all its possibilities because only one measurement result actually happens. Therefore at some point between its creation in the quon gun and its registration as an experimental result, a quon must repudiate the universal law of motion, halt its unbridled natural expansion, and contract into a single possibility corresponding to the single observed measurement result.

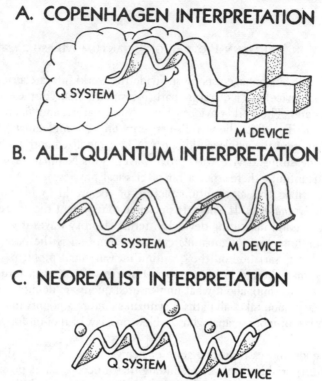

A. COPENHAGEN INTERPRETATION

Q SYSTEM

M DEVICE

B. ALL-QUANTUM INTERPRETATION

Q SYSTEM   M DEVICE

C. NEOREALIST INTERPRETATION

Q SYSTEM   M DEVICE

FIG. 8.3   *Three pictures of a quantum measurement. A. The Copenhagen interpretation envisions both the quantum system and measuring device as theoretically inaccessible. The quantum system's dynamic attributes have no definite values; the M device, on the other hand, is experienced by human observers "classically"—its attributes always appear to possess definite values. The wave function is a technical tool which expresses the relationship existing between these two radically different kinds of being. B. Von Neumann's all-quantum interpretation represents both system and M device as proxy waves. To make this description work, the system's waveform must "collapse" during a measurement. C. Neorealists envision everything (both system and M device) to be made of particles which interact with one another via invisible superluminal waves.*

Thus, according to von Neumann, a quon follows not one law of motion but two. Everywhere in the universe, proxy waves are expanding (called "Type I process" by von Neumann). However in the measurement act, and nowhere else, proxy waves contract to a definite result (a so-called "Type II process"). After contraction these entities are still proxy waves;

they never turn into classical objects. Consequently when one attribute's realm suddenly contracts, its realm of conjugate possibilities explosively expands.

Physicists call von Neumann's Type II process "the collapse of the wave function"; it's also called the "quantum jump." Von Neumann's all-quantum description will not work unless such a collapse really occurs as a physical process in every quantum measurement. My sixth variation on the measurement problem is this: *How and where does the wave function collapse occur?*

WHERE DOES THE WAVE FUNCTION ACTUALLY COLLAPSE?

Von Neumann was understandably anxious to find a natural location for the wave function collapse, which is essential for his interpretation of quantum theory. He systematically examined the measurement process for clues for a special feature of measurement which might give rise to a Type II process. He visualized the measurement act as broken into small steps called the von Neumann chain, stretching from the quon gun to the observer's consciousness where the measurement result is ultimately registered. Each process in between is a link in von Neumann's chain. A solution to the measurement problem, according to Swiss chemist Hans Primas, would consist of "severing von Neumann's chain at the first true measurement act." In other words, where in fact is a quantum measurement actually accomplished?

While searching for a natural place to break his chain, von Neumann proved an important mathematical fact that deepens the mystery of measurement. Von Neumann showed that *as far as final results are concerned, you can cut the chain and insert a collapse anywhere you please.* This means that the results themselves can offer no clues as to where to locate the division between system and measuring device.

Von Neumann's version of the measurement problem reminds me of a cartoon by Sidney Harris, an artist who specializes in poking fun at scientific foibles. A mathematician has outlined his work on the blackboard for one of his colleagues. The argument consists of the usual incomprehensible symbols, but near the end there's a gap in which he's written, "And then a miracle occurs." The math then resumes and proceeds to its logical conclusion. The mathematician's colleague is pointing to the gap and saying, "I think you should be more explicit here in Step 2."

On each side of the wave function collapse, von Neumann erects impeccable mathematical structures familiar to quantum physicists—the world expressed as proxy waves. However, separating these two sides of the argument—the world unmeasured and the measured world—is a logic gap in which von Neumann effectively writes, "And then a miracle occurs."

Von Neumann could not find a natural place to locate his "miracle." Everything, after all, is made of atoms: there's nothing holy about a measuring instrument. Following the von Neumann chain, driven by his own logic, in desperation von Neumann seized on its only peculiar link: the process by which a physical signal in the brain becomes an experience in the human mind. This is the only process in the whole von Neumann chain which is not mere molecules in motion. Von Neumann reluctantly came to the conclusion (Quantum Reality #7) that human consciousness is the site of the wave function collapse.

This direct intervention of consciousness in every measurement is what I call "observer-created reality of the second kind" to distinguish it from the mild kind of observer-created attributes entailed by the quantum meter option. In von Neumann's consciousness-created world, things (or at least their dynamic attributes) do not exist until some mind actually perceives them, a rather drastic conclusion but one to which this great mathematician was forced by sheer logic once he decided to take the quantum measurement problem seriously.

Fig. 8.3 illustrates three major approaches to what goes on in a quantum measurement: 1. Bohr's Copenhagen interpretation divides the world into quantum and classical realms—both incomprehensible—whose relationship is represented by a fictitious proxy wave; 2. Von Neumann's all-quantum picture represents both quon and M device with proxy waves which are connected by the so-called wave function collapse; 3. David Bohm, Louis de Broglie and other neorealists describe the physicist's world—consisting of systems and M devices—as being made solely of particles connected by (superluminal) waves.

Each of these approaches to quantum measurement has its drawbacks; none gives a completely satisfactory picture of the measurement act. The Copenhagen interpretation endows the measuring instrument with magical properties—the ability to reduce possibility to actuality, for instance—while removing such instruments in principle from logical analysis. Von Neumann restores the measuring instrument to an equal status with the rest of the world but transfers its magical properties to a mysterious and elusive event, the wave function collapse. The neorealist model of reality sanctifies neither measuring device nor measurement act; neorealist measurements are just ordinary interactions. However, the price for this neorealist solution to the QMP is the necessary existence of invisible superluminal force fields.

Because of the unsatisfactory state of the measurement problem, many physicists have attempted to find solutions less drastic than those of Bohr, von Neumann, and Bohm.

DO QUANTUM ENTITIES POSSESS FUZZY ATTRIBUTES?

In the early days of quantum theory, Erwin Schrödinger wondered whether one might be able to devise an all-quantum world like von Neumann's with this difference: although measuring devices are fully quantum, their quantum effects are so small that for all practical purposes they act classically. Schrödinger's proposal depends on the fact that Planck's constant is so tiny that on the scale of ordinary objects it is effectively zero.

Schrödinger proposed that the attributes of measuring devices as well as atoms are intrinsically ill defined, but the fuzziness of measuring devices is so small (on the order of Planck's constant) that to human senses they appear to possess well-defined attributes. In other words, Schrödinger suggests that a single atom as well as an individual M device possesses not one momentum but a range of momenta—momenta that exist not merely in potentia but in actuality. Schrödinger's proposal seems plausible when one realizes how small Planck's constant actually is compared with ordinary acts.

The dimensions of the quantity "action" in physics is energy times time (erg seconds). One erg second is not very large—about equal to the amount of action in a blink of your eye. Planck's constant of action is incomprehensibly smaller than an eye blink. It is one erg second divided by a billion . . . divided by a billion . . . then divided once more by a billion. Planck's constant is about 1×10^{-27} erg seconds.

Astronomers reckon the whole starry universe to be approximately 10^{27} centimeters wide—the distance light has traveled since the universe began. One centimeter is about the size of a sugar cube. A line of sugar cubes stretching across the universe would contain about 10^{27} cubes. This row of cubes spans not just the solar system, not just our local galaxy or galactic cluster, but the entire physical universe—everything there is.

One eye blink contains as many quanta of action as there are sugar cubes in that long white line. So minuscule is the scale of quantum events compared to the actions of everyday life that it's a wonder humans ever found out about the quantum world at all. If, as Schrödinger conjectured, the attributes of measuring devices were fuzzy to the extent of a few quanta, this fuzziness would be utterly undetectable, like a firefly in the glare of the sun. For atoms whose entire beings are composed of just a few quanta of action, Planck's constant is a big deal. For macroscopic instruments with actions on the scale of eye blinks, a few quanta more or less make no noticeable difference.

Despite the fact that their fuzziness could never be directly observed, Schrödinger concluded that measuring devices cannot possess fuzzy attributes. He argued that even though M devices would have realms of possibility too small to measure, it's easy to imagine experiments which split these tiny realms into two disjointed domains, each of which has very different macroscopic consequences.

For instance, in our polarization experiment the realm of possibility of each D photon is exactly cleaved in two by the H/V crystal. This photon, after passing through the crystal, becomes a superposition of V and H photons which are spatially separated. According to the laws of quantum motion (expand to fill all possibilities) this state evolves into a state in which both the up and down counters are triggered at the same time by a single photon. In this experiment the fuzziness of the microworld spreads in such a way that it engulfs the whole apparatus: now two large pieces of phosphor, glass and metal are forced into a quantum superposition.

To dramatize his argument, Schrödinger expanded this measuring device to include a cat. Imagine, he says "a hellish device": the outputs of the up and down detectors go into a closed box in which an innocent cat dwells. If the up counter fires, the cat lives; if the down counter fires, the cat dies. Only one photon enters the apparatus. What happens to Schrödinger's cat?

Because of Schrödinger's assumption that macroscopic objects possess fuzzy attributes, the photon's two simultaneous paths lead to the counters' two simultaneous triggerings, which results in both an up and a down signal which makes the cat *both alive and dead at the same time!*

The photon's realm is only one quantum wide, but a crystal of calcite is keen enough to split it. This split spreads to the photon detectors and ultimately to Schrödinger's cat, which must now exist—if his conjecture is correct—in a peculiar state: half-dead cat/half-live cat. And as my colleague Bruce Rosenblum remarks, "This does not mean the cat is *sick*."

Unless our perceptions are terribly mistaken, it is absurd to attribute such a bizarre form of existence to a cat or to a measuring device: nobody has ever seen a superposition of up/down measuring devices, let alone live/dead cats, and no one ever will. Schrödinger concludes that we are simply mistaken if we believe that measurement devices possess slightly fuzzy attributes. This attempted solution to the measurement problem is a dead-end street.

Somewhere between crystal and cat, the quantum rules have to change: the system's spread-out attributes must somehow turn into one unique actuality. Schrödinger's informal argument suggests what von Neumann proved more rigorously with his mathematics: if the whole world is described quantum-mechanically, in terms of proxy waves, then somewhere between the quon source and final result a "wave function collapse" must occur. If the wave function never collapses but always expands to fill its open possibilities, then Schrödinger's hellish device shows how easy it would be to produce cats that were both alive and dead at the same time.

Schrödinger's argument against fuzzy attributes depends on the fact that in this experiment we know (because each typed output mark is either up or down, never both) that the cat will either be alive or dead but not both. We assume here that as far as its macroscopic behavior goes, a cat is no different from a typewriter. Although we have good reason to believe that the cat always exists in a definite state of health in this experiment, quantum theory seems to permit in principle other experiments in which the outcome is not so clear-cut. Vladimir Braginski's sapphire bar, for instance, is larger than any cat and exists, after the Russians exercise their meter option in favor of attribute X_1, in a quantum state consisting of half momentum, half position waves summed with a definite phase: an unusual situation for a "classical" object. Perhaps we've never seen a live/dead "Schrödinger cat" only because we don't know how to look for one.

At a recent conference on quantum chemistry, physicist Frederik Belinfante, author of an important survey of hidden-variable models of reality, commented that if quantum theory permitted Schrödinger catlike phenomena to happen, one could not only kill cats with quantum machines but also bring them back to life. He predicted that quantum theory might someday become an important branch of veterinary medicine.

From what we know about quantum theory, it's easy to see how to build a machine for reviving dead cats. All we have to do is precisely measure an attribute which is *conjugate* to the live/dead attribute; the uncertainty principle does the rest. Start with a cat in a well-defined dead state. Exercise your meter option in the following manner. Put this cat through a filter which only passes cats with a well-defined value of an attribute called "diagonal cat"—an attribute conjugate to the live/dead attribute. Half the (formerly dead) cats that pass such a filter will be alive. If your cat is still in the dead state after passage through the diagonal-cat filter, put him through again. This filter has only a 50 percent cure rate but it can be used over and over.

CAN PHASE RANDOMIZATION COLLAPSE THE WAVE FUNCTION?

In Schrödinger's experiment the photon takes both paths but the cat doesn't. What special feature of macroscopic objects prevents a cat from splitting into a superposition of possibilities? Some physicists believe that phase randomization is what separates cats from photons. Cats are quantum entities too, but their phases are mixed up; a photon's phase on the other hand is nice and orderly. Does the process of wave function collapse occur whenever a system's phases become sufficiently random?

It is not difficult to find mechanisms for phase randomization inside measuring devices. Italian physicists Antonio Daneri, Angelo Loinger, and Giovanni Maria Prosperi show that *the thermodynamics of large bodies* can randomize phases. Russian physicist Dmitri Blokhintsev shows that the process of *amplification,* which makes a quantum process visible to human eyes, will inevitably randomize quantum phases. Others blame phase randomization on the *irreversible process involved in making a record.* Physicists H. Dieter Zeh and Wojciech H. Zurek show that *interactions with the environment* are continually randomizing the phases of macroscopic objects. In their view, cats and photon counters don't split because the environment is always watching them.

Von Neumann proved that you could put the wave function collapse anywhere between source and observer without changing the results. This means that the collapse site cannot be located by appeal to experiment. However, if you put the collapse too close to the quon source you will spoil the results of *other* experiments which you could have done on the same quon. If you collapse the wave function prematurely, these other experiments will give the wrong results. The results of these other experiments allow us to exclude certain locations for the collapse site.

For instance, suppose we assume that the input D photon collapsed inside the analyzing crystal: instead of taking both paths *(V* and *H)*, it just takes one. This collapse location is consistent with the results that for each photon only one symbol (Up or Down) is typed on the output tape, but such a *premature collapse* will not agree with another experiment we can do which determines whether a photon takes one path or two.

Using mirrors, for instance, we can combine light from the two crystal channels and look for the *wavewise interference* of polarization attributes which is characteristic of quons that are taking both paths. If we see such interference, then we know the wave function has not yet collapsed. In this case, when we combine beams we see interference effects between polarization attributes: immediately after the crystal, the photon evidently takes both paths.

Let's follow the von Neumann chain a bit farther. Does the wave function collapse when the photon excites a phosphor molecule? Most physicists would argue that it does not. After the phosphor is excited it returns to its ground state: it does not make a record of the photon's presence. If we were clever enough, we could make the light from both the up and the down phosphors interfere, and verify that when it passed the phosphors the measurement process was still split.

What about the next step in the von Neumann chain? The phosphor light (still presumably split into two simultaneous possibilities) passes through a pair of focusing lenses. One well-known property of lenses and mirrors is that they preserve phases: if they did not, they could not form clear images. So it's unlikely that the wave function collapses at the lenses.

The light-sensitive screen is next in the von Neumann chain. Here light interacts with the electron wave in a silicon crystal, promoting it to a higher energy state. Since the up electron wave is in a physically different crystal from the down electron, it is difficult to get them to interfere. However, we could imagine doing the same experiment using a single slab of silicon instead of two separate detectors. Again in principle, we could observe interference effects between the two electron excitations in the silicon crystal.

It seems that if you are sufficiently ingenious you can push the site of the wave function collapse as close as you wish to the observer. However, this experimental method depends on the fact that the waves in each path have well-defined phases.

This particular test for premature wave function collapse depends on the experimenter's ability to bring the split beams together to show an interference pattern. But we saw in Chapter 5 that when waves add *with random phases* their interference pattern disappears. So if phases become random somewhere along the line, then we cannot apply this test for wave function collapse. This does not in itself mean that once the waves become random, they have collapsed; it means only that this way of testing for premature collapse will not work anymore.

Besides making it difficult to test for the presence of simultaneous possibilities, the randomization of phases has another peculiar consequence for quantum theory which we can understand by looking at what phase randomization does to ordinary waves. We recall from Chapter 5 that ordinary waves added with definite phases do not conserve energy (amplitude squared) everywhere, but show local regions of energy surplus and deficit. However, when the waves are randomized energy is conserved everywhere: energy in one beam adds like ordinary arithmetic to energy in the other beam.

The same thing happens to randomized quantum waves, but here *probability* (quantum amplitude squared) takes the place of energy in ordinary waves. Quantum waves added with definite phase do not conserve probability everywhere, but show local regions of probability surplus and deficit (interference of attributes). However, when these waves are randomized, probability is conserved everywhere: probability in one beam adds like ordinary arithmetic to probability in the other beam. In other words, when a quantum wave's phase is randomized, its corresponding probabilities combine exactly like classical dice probabilities.

This remarkable fact, that after phase randomization quantum probabilities behave *numerically* the same as classical probabilities, is used by some physicists to argue that once phase randomization has occurred, quantum and classical probabilities are *conceptually* the same. That is, they claim that randomization of phase by itself is sufficient to bring about wave function collapse and convert a situation in which a quon take both paths (quantum ignorance) to a situation in which it takes only one (classical ignorance).

Although phase randomization can certainly scramble paths, it is difficult to see how it can destroy them: the law of the realm guarantees that a quon's possibilities can never be reduced below an action quantum. Just as no amount of mixing will turn black sand and white sand into gray sand, so no amount of phase randomization will turn two paths into one. Whenever one looks closely at claims that randomization by itself collapsed the wave function, one always finds that the collapse—the conceptual transition from quantum to classical ignorance—had to be put in "by hand." Although phase randomization may muddy the waters, it cannot hide the fact that there is always a place in the analysis where one simply admits, "And then a miracle occurs." Phase randomization is evidently present in all measurement situations but by itself does not constitute a "measurement." The scrambling of quantum phases seems to be a necessary but not sufficient condition for wave function collapse.

Although phase randomization can destroy some aspects of wave behavior and mimic to some extent the behavior of classical probability, certain aspects of wave behavior are immune to its effects. We saw for instance in Chapter 5 that when the Airy pattern was randomized, interference was destroyed but diffraction survived. The spectral area code (Heisenberg uncertainty principle) is another wave property that remains valid whether phases are orderly or not. Perhaps there are other innate wave properties yet to be discovered which survive phase randomization and whose quantum analogs may help us penetrate deeper into the mystery of the *collapse of the wave function:* where the strange world of the quantum quietly turns into the world of everyday life.

Appendix: Calcite, a Crystal That Splits Photons

At the time of Newton, a sailor discovered in Iceland a transparent mineral with remarkable optical properties. Iceland spar, more commonly called calcite, is birefringent: It bends light along two different paths depending on its polarization.

This crystal's sensitivity to polarized light suggests that the mysterious "sailing stone" mentioned in the Icelandic sagas may have been a species of calcite. As a direction finder, lodestone would be useless so close to the Earth's magnetic pole. Sailors in such high latitudes probably steered by the stars and sun. But during the long summer twilights the sun is often out of sight below the horizon. Because light from the sky is partially polarized in a particular sun-centered pattern, Viking sailors could have fixed the sun's direction at sea by observing skylight polarization through a calcite crystal. Although we will probably never know for sure, this crystal, so useful in modern reality research, may long ago have helped Norsemen discover America.

When a beam of light goes into a calcite crystal, two beams come out. One of these beams is called the *ordinary ray* because it obeys the conventional laws of optics. These laws say, for instance, that a beam of light striking a transparent surface head-on does not bend. Calcite's second beam is named the *extraordinary ray* because it flouts these optical regulations. For instance when light strikes calcite head on, the ordinary ray goes straight but the extraordinary ray bends.

This strange schizophrenic behavior of light in calcite challenged the best minds of the seventeenth century. Dutch physicist Christian Huygens gave the first scientific explanation of this marvelous crystal. Huygens analyzed calcite light by conceptually breaking it up into little spherical wavelets, a technique which inspired Feynman's sum-over-histories version of quantum theory three centuries later.

The key to calcite's behavior is its *optic axis*—a special direction, indicated by an arrow in Fig. 8.1, that runs through the crystal. Light polarized parallel to this axis travels through the crystal normally. Light polarized at right angles to the optic axis takes a deviant route—the extraordinary way. If the optic axis is oriented vertically, vertically polarized photons take the ordinary path (up), horizontally polarized photons go the extraordinary way (down).

In the modern view, calcite cleaves light in two because its crystal structure is sensitive to the difference between the quantum waveforms associated with the polarization attributes. According to quantum theory, the calcite crystal is a window into the microcosm: its double beam is indicative of the two-valuedness of the photon's polarization attribute.

9

Four Quantum Realities

Physics takes its start from everyday experience, which it continues by more subtle means. It remains akin to it, does not transcend
it generically; it cannot enter into another realm. Discoveries in
physics cannot in themselves—so I believe—have the authority of
forcing us to put an end to the habit of picturing the physical
world as a reality.

Erwin Schrödinger

Quantum theory works like a charm: it correctly predicts all the quantum
facts we can measure plus plenty that we can't (such as the temperature of
the sun's interior) or do not care to (the electron's "piano attribute," for
instance). This theory has passed every test human ingenuity can devise,
down to the last decimal point. However, like a magician who has inher-
ited a wonderful magic wand that works every time without his knowing
why, the physicist is at a loss to explain quantum theory's marvelous success.

What does it mean to "explain" a theory? Just imagine what one would like to know about the magician's wand, namely the hidden reality responsible for its magical operation. Quantum theory is more than a lucky gift out of the blue; this theory's unprecedented predictive power suggests that it makes contact with some real features of the physical world. An "explanation" of quantum theory would tell us what sort of world we live in that allows such a curious wave-mathematical technique to foretell this world's gestures in such precise detail.

Quantum theory resembles an elaborate tower whose middle stories are complete and occupied. Most of the workmen are crowded together on top, making plans and pouring forms for the next stories. Meanwhile the building's foundation consists of the same temporary scaffolding that was rigged up to get the project started. Although he must pass through them to get to the rest of the city, the average physicist shuns these lower floors with a kind of superstitious dread. New York University professor Daniel Greenberger, speaking at a recent *Festschrift*, speculated about why most physicists avoid the quantum reality question:

"This sudden success on a grand scale, after a generation of desperate striving by great minds, lends a heroic, even mythic, quality to the history of [quantum theory]. But, inevitably, it has also led to a sensitivity on the part of physicists, a kind of defensiveness, ultimately arising from the fear that the whole delicate structure, so painstakingly put together, might crumble if touched. This has tended to produce a 'Let's leave well enough alone' attitude, which I believe contributes to the great reluctance most physicists have to tinker with, or even critically examine, the foundations of quantum theory. However, fifty years have gone by and the structure appears stronger than ever."

Physicists' reality crisis consists of the fact that nobody can agree on what's holding the building up. Different people looking at the same theory come up with profoundly different models of reality, all of them outlandish compared to the ordinary experience which constitutes both daily life and the quantum facts. Physicists differ over which parts of this theory they will take seriously and which parts they will ignore as empty formalism having no counterpart in the real world. Which different picture of quantum reality you end up with depends on what parts of quantum theory you take seriously. In this chapter and the next I examine how the eight major quantum realities arise from the selective emphasis of certain features of quantum theory and the neglect of others.

Quantum Reality #1: The Copenhagen interpretation, Part I. (There is no deep reality.) The Copenhagen interpretation, developed mainly by

Bohr and Heisenberg, is the picture most physicists fall back on when you ask them what quantum theory means. Copenhagenists do not deny the existence of electrons but only the notion that these entities possess dynamic attributes of their own. Although an electron is always *measured* to have a particular value of momentum, it is a mistake, according to Bohr, to imagine that *before* the measurement it possessed some definite momentum. The Copenhagenists believe that when an electron is not being measured, it has no definite dynamic attributes.

Quantum theory was developed almost solely by Europeans. J. Robert Oppenheimer, one of the few Americans to have participated in Bohr's Copenhagen Institute, here explicitly denies the existence of the major attributes with which classical physics described a particle's external motion: "If we ask, for instance, whether the position of the electron remains the same, we must say 'no'; if we ask whether the electron's position changes with time, we must say 'no'; if we ask whether the electron is at rest, we must say 'no'; if we ask whether it is in motion, we must say 'no.'"

Some physicists confuse the Copenhagen doctrine with a *pragmatic* interpretation of quantum theory. The pragmatist regards any theory as a mere mathematical machine for generating numbers which he then compares with experiment. A pragmatist is concerned with results, not reality. The pragmatist refuses on principle to speculate about deep reality, such a concept being meaningless from his point of view. Pragmatism is an intellectually safe but ultimately sterile philosophy.

A pragmatist would refuse on principle to comment on the existential status of an unmeasured electron's attributes. No timid pragmatists, these students of Bohr! The Copenhagenists claim not that such attributes are meaningless but that they are nonexistent. They base their conclusions about an unseen quantum reality not on some abstract philosophical principle applicable in all cases but on the specific structure of quantum theory itself. Some theories of the world (Newtonian mechanics, for instance) allow us to believe or not that unobserved entities possess their own attributes. Quantum theory, according to the followers of Bohr, does not permit us this option.

Copenhagenists take as the central clue to the nature of the quantum world the uncertainty principle, which limits humans' abilities to probe the microcosm. The uncertainty principle is more than just an irreducible fuzziness existing "out there." It seems to be tightly bound up with the process of measurement. There's no attribute, for instance, that is intrinsically uncertain; any attribute we please can be measured with perfect accuracy. However, our choice of what we will precisely measure makes conjugate attributes maximally uncertain. Quantum uncertainty is not tied to one particular attribute but slides from attribute to attribute as we change our minds about what to measure.

In the early days of quantum theory this measurement-dependent uncertainty was attributed to an unavoidable disturbance of the quantum system by measurement—a disturbance which could neither be minimized (because Planck's constant enforces a minimum action exchange) nor calculated (because of quantum randomness). However, the slippery, shifting uncertainty of conjugate attributes is much too systematic to be explained by simple disturbance models of quantum measurement.

A second argument against the disturbance model of measurement is the existence of "Renninger-style measurements"—measurements in which information is gained about a system through *the absence of a detection event*. Mauritius Renninger was a German physicist who first pointed out the possibility of gaining information by observing that "nothing happens."

As an example of a Renninger-style measurement, consider a quantum system that possesses just two possibilities—for instance, light from a distant star which can bounce off either mirror A or mirror B of a stellar interferometer. Suppose we know that a photon is on its way to the mirrors but do not know which path it will take. We place a detector in path A and it does not click. The absence of a click in path A tells us that the photon must have taken path B. We have measured the position of a photon without explicitly interacting with it. Renninger and others looked closely at such null measurements and found to their surprise that they are equivalent in every way to measurements in which something actually happens—that is, in von Neumann's terms they "collapse the wave function" as well as increase the uncertainty of the conjugate attribute.

In the disturbance model, an electron actually possesses attributes which are unpredictably changed by the measuring device. What goes on in a quantum measurement is not so simple, according to the Copenhagen school. Bohr's explanation of the slipperiness of quantum attributes is that such attributes do not belong to the quon itself but reside in "the entire measurement situation"—a phrase Bohr was particularly fond of. When we measure a certain attribute, we should not imagine that the electron actually possesses this attribute. Electrons possess no attributes of their own. An electron's so-called attributes are really relations between the electron and its measuring device and do not properly belong to either.

Bohr's notion of *relational reality* explains why an attribute's uncertainty depends on the type of measurement performed and why attributes are affected by measurements as non-disturbing as Renninger's "no-show" techniques. Since the electron's "attributes" reside in the *relation* between quon and M device, one can influence the attribute by merely changing the device.

Support for Bohr's relational reality concept comes from the way quantum theory actually works. A quantum measurement corresponds to analyzing a particular waveshape into a certain family of waveforms—tuba or piano waves, for instance. Each waveform family represents a particular quantum attribute. The question is, do attributes of this kind truly belong to the quon in question or do they partly belong to the M device—to the analyzer prism, for instance?

Record on tape a few minutes of traffic noise. If we analyzed this noise into tuba waveforms we could say that in a certain sense traffic noise is a chorus of tubas. But we could equally well say that it's a chorus of pianos or saxophones. There is a sense in which noise is *made of* these waveforms. However, we know that real traffic noise has no intrinsic tuba, piano, or saxophone qualities. These musical sounds derive as much from the analyzer as they do from the noise itself.

It is the same with quantum entities, says Bohr. Electrons do not possess position, momentum, or any other dynamic attributes. These so-called attributes are not intrinsic properties of quantum systems but manifestations of "the entire experimental situation." It makes no sense to talk about the dynamic attributes of the electron itself. Such "attributes" only arise in a particular measurement context and change when that context changes. According to Bohr, "Isolated material particles are abstractions, their properties being definable and observable only through their interaction with other systems."

An obvious feature of the ordinary world is that it seems to be made of objects. An object is an entity that presents different images from different points of view and to different senses, but all these images can be thought of as being produced by one central cause. No one sees (smells, feels) the same subjective apple, but everyone agrees that there is one objective apple that is the source of these varied sensations. Its division into objects is a most important aspect of the everyday world. But the situation is different in the quantum world (which is, after all, only the ordinary world examined closely).

The separate images that we form of the quantum world (wave, particle, for example) from different experimental viewpoints do not combine into one comprehensive whole. There is no single image that corresponds to an electron. The quantum world is not made up of objects. As Heisenberg puts it, "Atoms are not things."

This does not mean that the quantum world is subjective. The quantum world is as objective as our own: different people taking the same viewpoint see the same thing, but the quantum world is not made of objects (different viewpoints do not add up). The quantum world is objective but objectless.

An example of a phenomenon which is objective but not an object is the rainbow. A rainbow has no end (hence no pot of gold) because the rainbow is not a "thing." A rainbow appears in a different place for each observer—in fact, each of your eyes sees a slightly different rainbow. Yet the rainbow is an objective phenomenon; it can be photographed.

For Bohr, the search for deep reality—for "real" quantum attributes which an electron possesses independently of observation—is as misguided as looking for the rainbow's end. An electron's attributes do not belong to the electron itself but are a kind of illusion produced by the electron plus "the entire experimental arrangement."

We like to think that although the rainbow is not an object, it is really made of objects: an illusion constructed of non-illusory sunlight and rain. Likewise we need some hard facts out of which to construct the electron's illusory attributes. In the Copenhagen interpretation, the ultimate "hard fact" is the measurement device itself. In an obvious way, phosphor screens are made of electrons; the Copenhagenists contend that in a less obvious way, electrons are also made of phosphor screens.

In the Copenhagen interpretation there is a sense in which the world is not made of atoms but of M devices. Berkeley physicist Henry Stapp comments on this peculiar reversed order of reality:

"Scientists of the late twenties, led by Bohr and Heisenberg, proposed a conception of nature radically different from that of their predecessors . . . Their theoretical structure did not extend down and anchor itself on fundamental microscopic space-time realities. Instead it turned back and anchored itself in the concrete sense realities that form the basis of social life. This radical concept, called the Copenhagen interpretation, was bitterly challenged at first but became during the thirties the orthodox interpretation of quantum theory, nominally accepted by almost all textbooks and practical workers in the field."

Bohr treats M devices in a special way. He does not represent an M device as a possibility wave but considers it a solid actuality. By objectifying the M device, he can account for the "ordinariness" of quantum fact and avoid such monstrosities as Schrödinger's live/dead cat which arise if you believe that the same quantum rules hold for cats and electrons. Throughout his career Bohr continued to emphasize the "classical style" of existence enjoyed by ordinary objects. For instance:

"Even when the phenomena transcend the scope of classical physical theories, the account of the experimental arrangement and the recording of observations must be given in plain language, suitably supplemented by technical terminology."

Bohr recognizes here that the *form* of every quantum fact is identical to the form of every prequantum fact—that is, nothing special. It's an unchangeable fact of life that our direct experience of an electron (flash-on-a-screen) is no more mysterious than our direct experience of cats and rainbows.

But how can the measuring device evade the quantum rules which hold for every entity in the world? I think that Bohr would answer this question by saying that if the M device should become itself the object of measurement, then it would certainly have to obey the quantum rules, but then M device #2 must reside in a classical-style world.

However, we could imagine a third M device that measures device #2. Now device #2 must be represented by a possibility wave, and obeys the superposition and uncertainty principles, but device #3 dwells in the classical world. This endless procession of measuring devices measuring one another is called "von Neumann's paradox of infinite regress." Von Neumann's paradox results from the assumption of special non-quantum entities—measuring devices—while at the same time we know that such devices cannot really be special.

One of the drawbacks of the Copenhagen view is that it assigns a privileged role to measuring devices, describing them in terms of definite actualities, while every other entity is represented by superpositions of possibilities. Surely the world itself is not so divided but consists of a single reality. Another conceptual weakness of the Copenhagen interpretation is that it regards both the M device and the measurement act as ultimately unanalyzable. Thus, in the Copenhagen view quantum theory can explain with great exactitude the behavior of atoms, but is powerless to cope with the attributes of cats and apples in their roles as unscrutinized parts of "the entire experimental situation".

Quantum Reality #2: The Copenhagen interpretation, Part II. (Reality is created by observation.) Expanding on the Copenhagen interpretation's special role for M devices, Quantum Reality #2 emphasizes the observer's special status in a quantum world. Of quantum theory's many elements, the observer-created reality school stresses the quantum meter option—the observer's ability to intervene in reality by freely selecting which attribute he wants to look at. In waveform language, the meter option amounts to freedom to choose the waveform alphabet in which an entity's proxy wave will be expressed: whether the traffic noise shall be made from piano or tuba (or whatever) waves. By your choice of what attributes you look for, say believers in observer-created reality, you choose what attributes a system will seem to possess.

Professor John Archibald Wheeler has built the Institute of Theoretical Physics at Austin, Texas, into the world's most important center for quantum reality research—a sort of second Copenhagen in the Lone Star state. Wheeler has at various times championed several different quantum realities, but none so consistently as observer-created reality. "No elementary phenomenon is a real phenomenon until it is an observed phenomenon," Wheeler maintains. Perhaps this quantum reality should be called the "Austin interpretation of quantum theory" in honor of Wheeler's institute.

If one accepts that common phenomena like the rainbow are observer-created, one should not be so surprised by such claims made on the electron's behalf. An electron after all is surely stranger than a rainbow. Wheeler takes observer-created reality a step beyond rainbows with what he calls a "delayed-choice experiment." In such an experiment, the observer creates not only present attributes of quantum entities, but also attributes that such entities possessed far back in the past, which by conventional thinking existed long before the experiment was conceived, let alone carried out. The concept of a delayed-choice experiment is best illustrated by one of Wheeler's thought experiments: the gravity-lens interferometer. According to Einstein, gravity is a curvature in space-time which shows up, for instance, in the deflection of starlight grazing the sun.

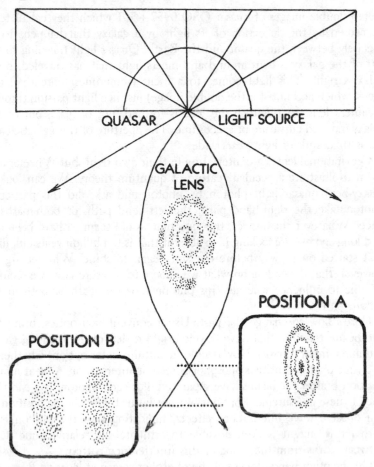

FIG. 9.1 *Galactic lens interferometer—an example of a "delayed-choice" experiment. Depending on whether the experimenter on Earth places his phosphor screen at A or at B, he seems to be able to make the photon travel one path around the galaxy (option B), or both paths (option A), changing a condition right now which by conventional thinking would seem to have been settled billions of years ago.*

The more massive an object, the more it bends space-time, consequently the more it deflects a beam of light. Since a typical galaxy contains some hundred billion suns, its ability to bend light is appreciable.

A quasar is an intense source of light of controversial origin which ignited shortly after the universe began. Recently astronomers have detected double images of quasar QSO 0957+561 which they calculate to be caused by the deflection of its light by a galaxy that happens to lie precisely between the quasar and the Earth. Quasar light traveling to the left of the galaxy is bent around and meets light that has traveled to the galaxy's right. This light comes to a focus (sometimes more than one focus, which accounts for the double image) just like light passing through a camera lens. In this case the lens is formed, not of glass, but of the galaxy-induced curvature of space-time. The aperture of this galactic camera is thousands of light years wide.

Gravitational lenses are interesting in their own right, but Wheeler uses them to illustrate a peculiar feature of quantum theory. We can look at galaxy-bent quasar light photon by photon and ask, did this particular photon take the right-hand path, the left-hand path, or both paths at once? Whatever the answer, this question would seem to have been settled long ago: we are looking here at light that is ten billion years old, light that started on its way before our sun began to shine. Wheeler argues, however, that depending on what we choose to measure now, we seem to be able to influence whether this photon took one path or both in the distant past.

Here's how to change the past. Using conventional optics, bring the two beams together that have traveled right or left of the focusing galaxy and allow them to cross. Now decide (quantum meter option) whether to put your photon-sensitive phosphor screen at intersection A or at a later position B after the beams have separated. For each photon, you can take one of these measurement options but not both. If you choose position A, you observe wave interference effects, indicating that the photon took both paths, and it is even possible to estimate the relative time delay between paths from the shape of this interference pattern.

On the other hand, if you put the phosphor screen at location B you see that each photon takes only one path, either left or right of the galactic lens. In our imagination we can picture this particular photon traveling partnerless for eons, because today we chose to do experiment B instead of experiment A. You can make the next photon take both paths by quickly choosing the other measurement option. By our choice of what we look at today we seem to be able to change a photon's attribute acquired billions of years before we were born.

Wheeler's delayed-choice experiment seems to show that the past is not fixed but alters according to present decisions. Popular philosopher Alan Watts reports that certain Eastern philosophies have come to a similar conclusion concerning the creative power of the present tense: "The moment of the world's creation is seen to lie, not in some unthinkably remote past, but in the eternal now."

Note that our ability to change the past is limited. We can indeed choose whether each quantum entity becomes a one-path or a two-path photon, but we cannot select where the two-path photon will fall in the interference pattern or which path the one-path photon will take. Such details are completely outside anyone's control or prediction: we call them "quantum random." Observer-created realities based on the meter option (observer creation of the first kind) can select the type of attribute a quon shall possess but not its particular *attribute value*.

The attributes of photons and electrons may well be observer-created, but what about the attributes of oranges and apples? Some adherents to observer-created reality, invoking arguments similar to Schrödinger's on behalf of his cat, believe that only quantum entities such as electrons and photons possess observer-malleable attributes. Wheeler, for instance, stresses that only *elementary phenomena* are unreal until observed. Presumably the attributes of non elementary phenomena such as cats and apples are real whether we look at them or not. However, it is difficult to draw a sharp line between classical and quantum entities. Other physicists of OCR persuasion draw no such line: for instance, Cornell's N. David Mermin believes that the attributes of all entities—cats, oranges, rainbows, even the moon and stars—are not real until somebody looks at them under the auspices of a particular meter option.

In addition to differing over which entities can benefit from an observation, partisans of an observer-created reality do not agree on what counts as an observation. Quantum theory itself does not say what is or is not a measurement but tells us that if we can find out how to make a measurement, it will predict the results. Luckily we already knew how to make measurements before the advent of quantum theory.

Wheeler and many of his colleagues have concluded that the essence of measurement is *the making of a record*. In their opinion, a Geiger counter or Polaroid (self-developing) film is competent to act as observer, assigning attributes via the meter option to observed entities now and forever afterward in the past.

A few physicists believe that record-making machines are not enough: only a *conscious observation* counts as a measurement (observer-creation of the second kind). Until conscious observers came upon the scene, the universe existed in an indefinite state, unable to decide even what kinds of attributes it possessed let alone their particular values. Large portions of the universe (everything that's not being looked at right now by a conscious observer) are still in this indecisive situation, waiting for a conscious observer to grant them a more definite style of existence.

Bishop Berkeley taught that matter possessed reality only insofar as it was perceived by some mind. No believer in observer-created reality, even the most extreme, goes as far as Berkeley. Every physicist upholds the absolute existence of matter—electrons, photons and the like—as well as certain of matter's static attributes. However, observer-created reality physicists do believe that *dynamic attributes*—position and momentum, for instance—do not exist until they are actually observed. Electrons certainly exist—with the same mass and charge whether you look or not— but it is a mistake to imagine them in particular locations or traveling in a particular direction unless you actually happen to see one doing so.

Quantum Reality #3: Reality is an undivided wholeness. The contention of David Bohm and others that despite its obvious separations the world is a seamless whole is related to Bohr's notion that quantum attributes are not localized in the quon itself but reside (like the position attribute of a rainbow) in "the entire experimental arrangement." Certain features of quantum theory imply that this innocent expression "entire experimental arrangement" may have to include not only activities in the immediate vicinity of the quon's actual detector but actions arbitrarily remote in time and space from the detection site. Ultimately the whole universe may be implicated in a simple measurement, in the selection of a single quon's observed attributes.

The basis for this claim of undivided wholeness is rooted in a curious feature of quantum theory called "phase entanglement." All previously mentioned puzzles of quantum theory concerned the process in which *a single quon* acquires its attributes. The concept of "phase entanglement" arises when we consider how *two or more interacting quons* acquire their attributes.

Whenever two quons meet, so do their representative proxy waves. Their time together is represented by a merging of proxy waves. The separation of these quons is represented by a separation of proxy-wave amplitudes, but the phases of the two quons do not come apart. Instead these phases become entangled in such a way that interference effects at quon A depend instantaneously on the disposition of quon B.

The reason that quantum waves become phrase-entangled and ordinary

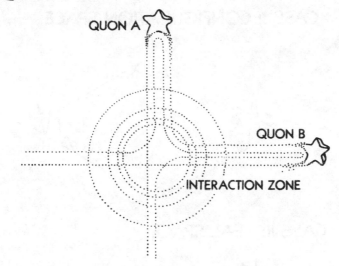

Fig. 9.2 *Phase entanglement. When two quons meet and separate, their amplitudes pull apart but their phases do not. All of quon A's phases depend on quon B's situation and vice versa. Phase entanglement mixes the phases internal to A and B as well as the relative phases between A and B.*

waves don't is that quantum waves do not make their home in ordinary three-dimensional space but in a place called *configuration space*. The difference between a couple of waves in ordinary space and in configuration space is illustrated in Fig. 9.3. For simplicity we imagine both waves to be one-dimensional, like waves on a string. In ordinary space we see two waves, A and B, traveling on the same one-dimensional string. The same situation represented in configuration space looks like one wave in two dimensions: waves A and B are not separate but different aspects of a single waveshape.

Configuration space consists of *three dimensions for each quon*. Thus the proxy wave for the two-quon hydrogen atom (electron plus proton) resides in a space of six dimensions. The main reason that physicists consider the wave function to be fictitious is that it moves around in a space with many more dimensions than our own.

A disturbing feature of phase-entangled quons, first emphasized by Erwin Schrödinger, is the strange action-at-a-distance such entangled systems seem to possess—at least on paper. Because of their phase-connectedness (or alternatively because they are represented by a *single wave* in

CASE I: CONFIGURATION SPACE

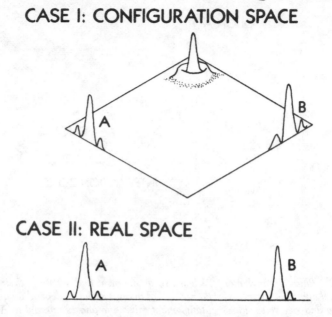

CASE II: REAL SPACE

FIG. 9.3 *Configuration space versus real space. Two waves (A and B) are traveling on a one-dimensional string. In Case I they are described as partial aspects of a single wave in two-dimensional space. In Case II they are described as two separate waves on the same string. The quantum proxy wave for two quons moves not in real space (Case II) but in configuration space (Case I)—quantum possibilities develop in a conceptual arena with many more dimensions than three.*

multidimensional space) an action on quon A seems to have an instantaneous effect on the wave function of quon B even when the two quons are no longer capable of interacting via conventional forces. Schrödinger found this apparent instant connection so unusual that he called it not *one* but *the* chief feature of quantum theory.

However, the fact that such a tight connection between separated quons exists in the *formalism* of quantum theory is no guarantee that such a connection exists *in reality*. Quantum connectedness may be as fictitious as the dotted lines that outline the constellations in star maps. Despite the fact that "something" seems to be linking two quons instantly (faster than light), quantum theory does not seem to permit any messages to be sent along this link (see Eberhard's proof in Chapter 13), a fact which heightens the impression that such links are spurious.

The fact that observing a quon *here* instantly changes the wave function *there* (where "there" may be billions of miles away) is another good argument for the fictitious nature of the proxy wave. If the wave function were real, it would have to change its shape drastically over large distances at faster-than-light speeds. As far as we know, nothing real can move that fast. To cope with such objections, Heisenberg suggested that the wave function does not represent a real situation but rather our knowledge of a physical situation. Wave function collapse is not an actual physical event but represents the change that occurs in our knowledge when we become aware of the result of a measurement. The knowledge interpretation also makes Renninger-style measurements (in which no actual detections take place) easier to understand: in certain situations the fact that "nothing happened" can increase our knowledge with no accompanying physical change.

If a friend in Texas seals a silver coin in one envelope and a gold coin in another and mails the envelopes to Tokyo and London, the instant you open your envelope in Japan you know the contents of my envelope in England. But opening your letter causes no physical change in England (faster-than-light or otherwise) but merely involves a change in your knowledge concerning something happening far away and outside your control. We certainly learn something about one coin by looking at the other. However, this knowledge is not physically transmitted from London to Tokyo but resides in the preestablished relation between the two coins. The silver and gold coins are *correlated but not connected*. No real link exists between them. Likewise, according to Heisenberg's knowledge interpretation, nothing real connects isolated far-flung phase-entangled quons. Instantaneous phase connections exist in the mathematics but not in the real world. You should not take them seriously.

Many entities that seem to exist are not part of the real world—for instance, the virtual image. A virtual image appears to be someplace where in fact it is not. A rainbow is a virtual image: it can seem to be located where there is neither rain nor sunlight. Alice's world on the other side of the mirror is likewise a virtual image. As far as the laws of optics are concerned, these images are just as real as your nose. They can be photographed, magnified, and reflected just like the image of any real object. But ordinary experience has stricter criteria for what's real and what's not than the laws of optics. As any child knows, there's nothing at all in back of a mirror.

On the other hand the silver/gold coin analogy, however persuasive, may not be an appropriate model for quantum correlations. Heisenberg did indeed suggest that the wave function collapse represented a change in the observer's knowledge rather than a real physical event, but he was careful to add that such knowledge was not of some preexisting reality. Knowledge of a quantum system is not attained by passive inspection but only by active intervention on the part of the observer. "What we learn about," says Heisenberg, "is not nature itself, but nature exposed to our methods of questioning."

Just as quantum ignorance is different from the classical kind, the "knowledge" that we gain in a quantum measurement is of a different sort from the knowledge we gain from opening an envelope. Perhaps we cannot so easily dismiss the reality of instant connectedness by appealing to ordinary knowledge models such as correlated coins.

This suspicion that mathematical phase entanglement is evidence for a real world quantum connection was strengthened by John Bell's discovery that *quantum correlations are too strong* to be explained by ordinary knowledge models of the silver/gold coin variety. Bell's theorem, reviewed in Chapter 12, shows that to account for the behavior of certain two-photon systems, extremely drastic models of reality must be invoked, models that necessitate the real existence of a pervasive and powerful long-range connectedness.

Quantum Reality #4: The many-worlds interpretation. This quantum reality, first dreamed up by Hugh Everett in 1957 while a Ph.D. candidate under John Wheeler, takes the quantum measurement problem seriously and solves it in a bold and flamboyant manner.

The measurement problem can be expressed in many ways. Everett saw it like this: the orthodox ontology treats measurement as a special kind of interaction, yet we know that *measurement interactions cannot really be special* since M devices are no different from anything else in the world. How, then, asks Everett, can we strip the measurement act of its privileged status and achieve within physics that *democracy of interactions* which certainly prevails in nature?

Bohr, for instance, assigns special status to measuring devices, conferring on them a classical-style actuality not possessed by the atomic entities under their scrutiny. Von Neumann, on the other hand, does not consider M devices special: he describes them in terms of possibility waves just like atoms. However, the price von Neumann has to pay to purchase this equality of being is the necessary elevation of the measurement act to special status. Unlike any other interaction in nature, measurement has the power to collapse the wave function from many parallel possibilities (the premeasurement superposition of possibilities) to just one (the actual measurement result).

Following von Neumann's picture of quantum theory, Everett represents everything by proxy waves, but he leaves out the wave function collapse. When a quantum system encounters an M device set to measure a particular attribute, it splits as usual into many waveforms, each corresponding to a possible value of that attribute. What is new in Everett's model is that correlated to every one of these system wave functions is a different M-device waveform which records one of these attribute values. Thus if the measured attribute has five possible values, the quantum-entity-plus-measuring-device develops into five quantum systems, each with a different attribute value paired with five measuring devices each registering that value. Instead of collapsing from five possibilities to one actual outcome, the quantum system in Everett's interpretation realizes *all five outcomes*.

To account for the stubborn fact that no one has ever seen one M device turn into five, Everett makes a not-so-modest proposal. The apparatus actually does split into five different parts, says Everett, but each part occupies *its own parallel universe*. A human being—one of Everett's critics, for instance—dwells in just one of these universes (at a time) and cannot perceive the other four. Likewise the inhabitants of the other four universes are not aware of their parallel partners.

The "ordinariness" of quantum facts in spite of the real existence of multiple universes is accounted for in Everett's model by the fact that each human observer perceives only a single universe. We do not know why human perception is limited to such a small sector of the real world, but it seems to be an unavoidable fact. We are not directly aware of these alternate worlds, but our own universe would not be the same without them.

Everett's quantum theory without collapse describes the world as a continually proliferating jungle of conflicting possibilities, each isolated inside its own universe. In that world (which we might call super reality) one M device splits into five. However, humans do not happen to live in super reality but in the world of mere reality, where only one thing happens at a time. We can picture Everett's super reality as a continually branching tree of possibilities in which everything that can happen actually does happen. Each individual's experience (lived out in mere reality, not super reality) is a tiny portion of a single branch on that lush and perpetually flowering tree.

All interactions in Everett's super-real world are of the same kind: two systems come together, get correlated, then start to realize all their mutual possibilities. A measuring device is just like any other quantum entity except that its macroscopic attributes happen to be especially sensitive to some attribute (usually position) of an atomic entity with which it may become correlated. Lots of entities become correlated with photons, but few qualify as photon detectors because their visible attributes are not significantly changed by this photonic association. Our phosphor/screen combination is different: it prints a mark on a tape whenever it correlates with a photon's position attribute. In Everett's model, M devices are not essentially different from anything else except in certain unimportant details.

Everett's many-worlds interpretation of quantum theory, despite its extravagant assumption of numerous unobservable parallel worlds, is a favorite model of many theoretical physicists because of all quantum realities it alone seems to solve the measurement problem with no arbitrary canonization of the process of measurement. In Everett's picture all measurement devices and measurement acts are fundamentally of the same nature as all other devices and acts. Strictly speaking, there are no "measurements" in the world, only correlations.

Einstein objected to suggestions of observer-created reality in quantum theory by saying that he could not imagine that a mouse could change the universe drastically simply by looking at it. Everett answers Einstein's objection by saying that the actual situation is quite the other way around. "It is not so much the system," Everett says, "which is affected by an observation, as the observer who becomes correlated to the system." The moral of Everett's tale is plain: if you don't want to split, stop looking at attribute-laden systems.

At a recent conference on the nature of quantum reality, Berkeley physicist Henry Stapp suggested an advantage that Everett's quantum reality confers on biological evolution and similar improbable but not impossible processes. Suppose, says Stapp, you could calculate the odds for life to begin on Earth and found them to be infinitesimally small but not actually zero. In the conventional single-universe model of things, something with a very small probability is effectively impossible: it will never happen. However, in the Everett picture everything that can happen does happen. If life on Earth is possible at all, then it is inevitable—in some corner of super reality. In Everett's bountiful multiverse, every little "could be," no matter how improbable, gets its time to shine.

10

Quantum Realities: Four More

I want to know how God created this world. I am not interested in this or that phenomenon, in the spectrum of this or that element. I want to know His thoughts, the rest are details.

Albert Einstein

Quantum Reality #5: Quantum logic. (The world obeys a non-human kind of reasoning.) In 1936 John von Neumann and Harvard mathematician Garrett Birkhoff proposed a new approach to quantum theory which they called quantum logic. An entity's "logic" means how its attributes combine to make new attributes.

The attributes of classical objects follow a familiar pattern called Boolean logic after George Boole, an Irish schoolteacher who first codified the structure of ordinary reasoning. Birkhoff and von Neumann show that because quantum attributes are represented by waveforms, they combine according to a peculiar "wave logic."

Consider a collection of entities that can possess attributes A or B or both. From the set of entities all of which possess attribute A and the set all of which possess attribute B we can form two new sets by combining sets A and B according to the logical operations AND and OR.

When we combine sets A and B according to the AND operation we get another set C (symbolically A AND B = C) which contains all entities possessing *both attribute A and attribute B*. The joint possession of both attributes may be regarded as a new attribute, attribute C. Note that A AND A = A.

Combining sets A and B according to the OR operation, we get another set D (symbolically A OR B = D) which contains all entities possessing *either attribute A or attribute B or both*. The possession of either of these attributes can be regarded as a new attribute, attribute D. Note that A OR A = A.

LOGIC LATTICES

To picture the structure of various logics, mathematicians use a diagram called a lattice which shows at a glance all AND/OR relations between attributes. A lattice orders a system's attributes according to their inclusiveness, with the most inclusive attribute on top. Attributes are connected by lines that show how each pair of lower attributes are connected by lines that show how each pair of lower attributes joins together to form a higher attribute. An example of a classical logic lattice is the *color lattice* which lays out the logical relations that exist among the primary and process colors (Fig. 10.1). Primary colors (red, green, blue) are the basis of *additive color mixtures* such as those which create full-color TV images. Color printing *(subtractive color process)* is based on mixtures of the process colors yellow, magenta, and cyan.

To compute the result of the AND operation on any pair of attributes A and B from a lattice diagram, find *the highest common attribute* that can be reached by *following lines downward* from A and from B. For instance, the operation (cyan AND magenta) is calculated by following lattice lines down from the cyan and magenta attributes. The highest attribute which these downward lines intersect is blue. We say that cyan AND magenta = blue which means that blue is the most inclusive attribute that cyan and magenta have in common. In like manner one can read off from the lattice the result of the AND operation applied to any pair of attributes.

To calculate the result of the OR operation on any pair of attributes A and B from a lattice diagram, find *the lowest common attribute* that can be

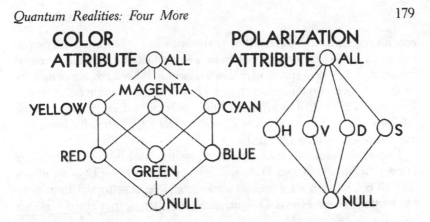

Fig. 10.1 *A classical and a quantum lattice. An attribute lattice displays in one diagram how any pair of attributes combines to make a new attribute. The color attribute lattice is Boolean; the polarization attribute lattice is not.*

reached by *following lines upward* from A and from B. For instance, the operation (cyan OR magenta) is calculated by following lattice lines up from the cyan and magenta attributes. The lowest attribute which these upward lines intersect is white. The white attribute is labeled "All" in Fig. 10.1 because white light is a mixture of all colors. We say that cyan OR magenta = white, which means that white is the smallest (least inclusive) attribute containing both cyan and magenta.

These AND/OR relations are more than abstract logical connections; they correspond to actual physical operations. The AND operation is carried out by putting together color filters. The operation cyan AND magenta = blue corresponds to putting a cyan filter (which passes only cyan light) together with a magenta filter (which passes only magenta light) to make a combination filter which passes only blue light.

Just as the AND operation corresponds to putting together two colored filters, so the OR operation corresponds to mixing together two colored lights. The operation cyan OR magenta = white describes the production of "white" light by mixing cyan and magenta lights. Since magenta and cyan both contain blue, the combination light is not pure white but bluish; nonetheless the light is still called "white" because a white-light detector (orthochromatic film or broad-band phosphor) will respond to such light over its entire range of sensitivity.

Color is a classical attribute which follows Boole's rules for AND and OR; the color lattice is an example of a Boolean lattice. A simple example of a non-Boolean quantum logic lattice is the lattice of polarization attributes. Polarization $P(\phi)$, introduced in Chapter 8, is a quantum attribute defined for any direction ϕ in space, which can take only two values. For simplicity we consider just two directions 0° and 45° with corresponding polarizations $P(0)$ and $P(45)$. Attribute $P(0)$ can take only the values H and V while attribute $P(45)$ takes only values D and S. The logic lattice for these two attributes is shown in Fig. 10.1.

Two of the quantum logical relations we can read from this lattice are H OR D = all and H AND D = null. The operation H OR D = all means that all polarization attributes whatsoever can be constructed from a superposition of just H and D light (provided that we put them together with the proper phase).

The operation H AND D = null means that H and D do not have a single polarization attribute in common: no conceivable light beam goes through both an H filter and a D filter.

Among the rules that every Boolean lattice must obey is the distributive law which says that for any three attributes A, B, and C, the following relation must be true:

$$A \text{ OR } (B \text{ AND } C) = (A \text{ OR } B) \text{ AND } (A \text{ OR } C)$$

Roughly speaking, the distributive law requires that the *parts* of a combination attribute—(B AND C) for example—can be acted upon, then combined, with the same result as acting upon the combination attribute alone. In other words, for systems which obey the distributive law, a combination attribute behaves like the sum of its parts.

Because it's Boolean, the color attribute lattice obeys the distributive law for all triplets of attributes, for instance:

$$R \text{ OR } (C \text{ AND } M) = R \text{ OR } B = M$$
$$(R \text{ OR } C) \text{ AND } (R \text{ OR } M) = W \text{ AND } M = M$$

Quantum logic lattices follow every Boolean rule with the exception of the distributive law. There are certain triplets of polarization attributes, for instance, for which the distributive law fails:

$$H \text{ OR } (D \text{ AND } S) = H \text{ OR } N = H$$
$$(H \text{ OR } D) \text{ AND } (H \text{ OR } S) = A \text{ AND } A = A$$

Evaluating the expression H OR (D AND S) directly, we get the attribute H. If we break D and S apart and calculate, we get attribute A. Since these two resultant attributes are not the same, the distributive law does not hold for this particular combination of attributes. Other examples can easily be found. The polarization attributes *P(0)* and *P(45)* form a non-distributive lattice. However, even though the polarization lattice violates one of the Boolean rules, it can always be divided into sublattices which are entirely Boolean. For example, the four-element sublattice consisting of attributes (H, V, N, A) and the sublattice consisting of (S, D, N, A) are both Boolean but the total lattice is not.

All of a classical object's attributes can be measured simultaneously: a classical object is completely open to view. A quantum entity is different: only a certain set of *compatible attributes* can be simultaneously measured. The Boolean sublattices of the full quantum logic lattice consist only of compatible attributes (H and V, for instance) that can be measured simultaneously. Compatible attributes follow ordinary logic—a logic which reflects the "surface ordinariness" of quantum fact—the human condition repeatedly stressed by Bohr that everything we experience in this world must be describable in "classical"—that is, Boolean language.

A quantum entity is never completely open to view: its visible compatible attributes represent only part of its full range of possibilities. The rest of the quantum lattice contains those hidden relationships which distinguish a quantum entity from a classical object. All quantum lattices consist of a union of Boolean sublattices (which some call the "isles of Boole") adrift in a wave-logical ocean of non-Boolean relations. Within each isle of Boole, normal logical relations prevail, corresponding to the surface ordinariness (Cinderella effect) of compatible quantum attributes. However, relations between the Boolean isles do not satisfy the distributive law, which suggests that for quantum entities something is fishy about the connection between the whole and its parts.

Quantum logic has made little impact on practical physics because most of the work carried out in its name has been concerned neither with the nature of reality nor the elucidation of experiments but with the mathematical study of non-distributive lattices for their own sake. A notable exception to this general preoccupation with formalism at the expense of physics is the work of Georgia Tech theorist David Finkelstein who hopes to use quantum logic to see beyond quantum theory into the actual processes that run the world.

Unlike Bohr, who held that quantum attributes are relational—shared jointly between quon and M device—or Wheeler, who believes that attributes are observer-created, Finkelstein takes the common-sense view that *quons actually possess their measured attributes* in the manner of classical objects. In Finkelstein's model, quons differ from classical objects not in how they possess their attributes but in *how these attributes combine to make new attributes.* Quantum entities possess classical attributes that obey a non-classical logic.

A classical object has parts that fit together in only one way—like clockwork which can be fully described by a parts list and an assembly drawing. A quantum entity, on the other hand, follows "wave logic" and doesn't have definite parts: it can break up in a great many different ways, as many ways as there are waveform alphabets into which its proxy wave can be analyzed. Each of these divisions yields a valid classical-style view of the quantum entity, but the connection between these ordinary views is governed by a wave logic whose image is not an assembly drawing but a non-distributive lattice.

THE THREE-POLARIZER PARADOX

As an example of wave logic in action, consider the so-called "three-polarizer paradox." A polarizing filter is a sheet of gray plastic that passes only one kind of polarized light.

Like the calcite crystal, a polarizing filter has an *optic axis,* and splits light into two beams—one beam polarized parallel to and one beam po-

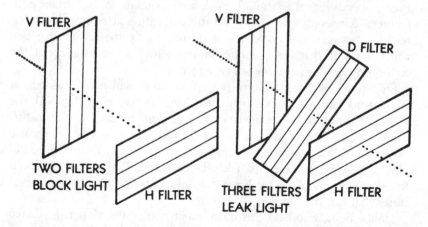

FIG. 10.2 *The "three-polarizer paradox." Crossed polarizers (H filter plus V filter) block all light. Adding a third filter (D filter) lets light through. Two absorbers suffice to stop the light. Why does adding a third absorber let light leak through?*

larized at right angles to its optic axis. Unlike the calcite that separates and transmits both beams, the polarizer *absorbs* the perpendicularly polarized beam and passes the beam polarized parallel to its optic axis.

We label a polarizing filter according to the type of light it passes. Thus a polarizer whose optic axis is horizontal is called a horizontal polarization filter (or H filter, for short). A polarizer is a general-purpose polarization analyzer: the same sheet of gray plastic can act as an H, V, D, or S filter depending on how it's tilted. In H-polarized light, an H filter acts like an open window. In V-polarized light, the same filter looks black.

Take an H filter and place it in front of a V filter. In this crossed-filter configuration, the light that's passed by the first filter (H light) is completely stopped by the second filter. Like a closed window, a pair of crossed polarizers absorbs all input light, no matter what its polarization.

Now take a third polarizer oriented in the D direction (diagonal = 45° to the vertical). This D filter absorbs S light (slant = −45° to the vertical) and passes D light. Put this D filter in front of the two crossed polarizers. Like any other kind of light going into the crossed polarizers, D light is completely absorbed. The three-polarizer stack remains opaque.

Now place the D filter *behind* the two crossed polarizers. Since no light passes the crossed polarizers, putting a filter in back of them has no effect. Again the three-filter stack blocks all light.

Now place the D filter *between* the two crossed polarizers. As the D filter moves into place, light begins to shine through the formerly opaque stack.

This is the three-polarizer paradox: together an H filter and a V filter block all light. Since any filter, however it may be oriented, does nothing but *absorb* light, how does the insertion of a third filter manage to let more light through? How can closing the window (adding another absorber) let in the light?

This strange behavior of light going through plastic sheets can be explained by representing the polarization attribute in the quantum manner as a particular waveform.

Two crossed polarizers pass no light because the light that goes through the first filter (H light) contains no V light, while V light is the only kind of light that passes the second filter.

The H attribute considered as a waveform can be analyzed into equal amounts of D light and S light added together with a particular phase. H polarization construed as a wavewise sum of D and S light may be symbolized as follows (where ⊕ is our symbol for wave addition):

$$H = D \oplus S$$

In turn D and S polarizations can be construed as a wavewise sum of equal amplitudes of H and V light added together with particular phases:

$$D = H \oplus V$$
$$S = H \oplus V$$

Both D polarization and S polarization are made from equal parts H and V light; the only difference is the phase with which these two lights are added. These expressions show that both D and S light consist of half V light. However, D and S add to make H light, which contains no V light whatsoever. The only way this can happen is if D's V light interferes destructively with S's V light:

$$H = D \oplus S = (H \oplus V) \oplus (H \oplus V) = H$$

destructive
interference

Thus although there is no V light at all in a pure H beam, such a beam contains waveforms S and D which are each half V light; however, this V light is entirely suppressed by destructive interference.

We now see how the absorption of an attribute can let through more light. The insertion of the third polarizer (D filter) absorbs S light out of the pure H beam. The removal of S light spoils the perfect destructive interference that yielded zero V light. The D light that's left is half V light with nothing around to cancel it. This uncanceled V light is what gets through the second filter. Putting in an absorber lets through more light. By removing one member of a pair of interfering attributes, the D filter appears to create light out of nowhere.

This curious behavior of polarizing filters is complementary to the behavior of Sheldon Glashow's quark-mediated K-particle decay discussed in Chapter 6. In Glashow's case the addition of a second quark channel completely suppresses a certain type of decay because the wave nature of the two channels permits them to cancel each other out. In the three-polarizer paradox, taking away light lets more light through (less is more); in Glashow's quark model of K-particle decay, adding a second decay channel suppresses decay altogether (more is less). As these examples show, quantum attributes do not obey the laws of ordinary arithmetic but a peculiar kind of wave arithmetic.

Quantum logicians explain the three-polarizer paradox as a simple case of non-distributive logic. The H filter lets through only photons that are H-polarized. The intervening D filter lets through photons that are both H-polarized AND D-polarized. However, even though no H photon can be V-polarized, some photons which are both H AND D happen to be V-polarized.

A NON-BOOLEAN ROUNDUP

The strangeness of this quantum-logical description of the three-polarizer paradox can be appreciated by applying it to a more familiar situation. I owe this illustration to peripatetic philosopher of science Ariadna Chernavska.

Suppose we pass cattle through a gate which only lets through horses and rejects all cows. Next we pass these horses through a second gate which lets through only black animals and rejects all white ones. Only animals which are both horses AND black can pass both gates. To our surprise, approximately half of such animals turn out to be cows!

Of course cattle don't behave this way, but if we believe the quantum logicians that's exactly what happens to polarized photons when they go through little sheets of plastic. A photon's attributes obey a non-human logic which we must learn to understand if we want to make sense of what's really going on in the quantum world.

Quantum Reality #6: Neorealism. (The world is made of ordinary objects.) The bottom line of many quantum experiments consists of a pattern of tiny flashes on a phosphor screen. Is it so obvious that such a simple phenomenon—the basis of all TV images—can be explained only by resorting to some bizarre quantum reality? Watching those little flashes of light appear on the screen one by one, it's easy to imagine that they are actually caused by little objects—by real electrons with position and momentum attributes all their own. This common-sense notion that the ordinariness of direct experience can be explained by an equally ordinary underlying reality is the basis for a quantum reality I call neorealism. Neorealists claim that the familiar objects that make up the everyday world are themselves made of ordinary objects; they believe, in short, that atoms are "things."

This straightforward view of the world's real nature has been generally dismissed by establishment physicists as misguided and hopelessly naïve. Werner Heisenberg, for instance, considered this way of thinking as outmoded as the idea of a flat Earth: "The ontology of materialism rested upon the illusion that the kind of existence, the direct 'actuality' of the world around us, can be extrapolated into the atomic range. This extrapolation, however, is impossible . . . Atoms are not things."

Not only was neorealism rejected by Heisenberg, Bohr, and other founding fathers of quantum theory as well as most of the scientific rank and file, it was condemned by the New Physics' foremost mathematical authority. World-class mathematician John von Neumann, in his quantum bible, considered the claims of the neorealists and conclusively rejected them. Von Neumann showed that because quantum theory represents attributes by waveforms, it makes predictions which no collection of ordinary objects can duplicate. In other words, if quantum theory is correct, neorealism is impossible. This conclusion, known as von Neumann's proof, strengthened the case for the prevailing Copenhagen view, considerably dampened physicists' enthusiasm for neorealist heresies, and effectively closed off research into object-based models of the world for more than twenty years.

QUANTUM MONTE

Though no one has ever seen an atom—our experience of such entities is tantalizingly indirect—von Neumann proved that whatever atoms may be, they cannot resemble ordinary objects. To get a feeling for how a mathematician could prove anything at all about an invisible reality, consider the following situation:

A fast-talking, flashy-dressed character on a New York street corner offers you a chance to earn some easy money. On his folding suitcase-table he lays out three cards face down. For a broker's fee of ten dollars he will give you two of these cards. If the cards you choose are the same color, he will buy them back for fifty dollars; a red and a black card are worthless. To prove this is not just a simple card trick, once you have made your choice he does not touch your cards but merely picks up the unchosen card and replaces it in the deck.

Suppose you decide to play his game a few times and don't win. You could blame that on bad luck. But after losing a dozen times, you might get suspicious.

"Say, there's something funny about this game."

"Why do you say that, buddy?" He grins, flashing a row of gold teeth.

"Because I can *prove* that no cards can do what these cards are doing."

"Really? I'd be very interested in such a proof," he replies. "I've wondered a lot about these cards myself."

"Look, you can't be dealing out three cards the same color, or I'd *certainly* get a pair. And if you deal out two cards of the same color plus one odd card, I'd have one chance in three of drawing a pair, but in twenty tries I haven't gotten any. I notice that nobody else gets any pairs either. And this is the clincher: *there's no other way to deal out three cards.*"

"Is that a fact?"

"Sure, don't you see? If you've got three cards and only two colors, at least two of the cards have got to have the same color. It's impossible to deal out three cards each with different colors!"

"Okay, kid, you got me. But you hadda be smart to figure it out. Only a guy with your kinda brains could have discovered these cards are special. I'll tell you my secret. They ain't ordinary cards; they're quantum objects my Uncle Johnny in New Jersey cooks up. Look, buddy, since you're in on the trick now, tell ya what I'm gonna do. I'll sell you the rest of this deck for the same price you paid for those cards you've got in your hand. As you can plainly see, you can make a fortune with quantum objects."

This hypothetical game (I call it quantum monte) behaves a lot like polarized photons. The arguments of the disappointed player follow the same logic as von Neumann's proof. If the three cards on the gambler's table are ordinary objects, there is no possibility for them all to be unmatched. Yet the player never gets a pair no matter how long he plays. *Hence the cards cannot be ordinary objects.*

This proof that the cards are not objects assumes that the gambler is honest. In the case of "quantum monte" this assumption may be a bit naïve. In the case of polarized photons, we can probably assume that the universe is not trying to fool us. As Einstein once remarked, "Nature is clever but she is not malicious."

This card game resembles a typical quantum experiment in more ways than one. Both quantum fact and quantum monte have the *element of chance* in common and the fact that each event that happens is, in itself, unremarkable (Cinderella effect). It's the *statistical pattern*, in both cases, that gives the game away.

Another way in which this game resembles the quantum world is that, in order to work, both must put a definite restriction on measurement. In quantum monte you cannot peek at the third card; in a quantum experiment you cannot look at the conjugate attribute. In both the card game and in the laboratory nothing is ultimately hidden: you can choose (meter option) to look at whatever you wish. However, every choice to look at one aspect necessarily entails not looking at something else.

DAVID BOHM'S ORDINARY-OBJECT MODEL OF THE ELECTRON

In the early fifties David Bohm published his popular textbook *Quantum Theory*, which remains even today one of the most lucid presentations of the Copenhagen view. While at Princeton, conversations with the aging Einstein, the most prominent agitator for a realistic world view ("Reality is the real business of physics.") weakened considerably Bohm's faith in Copenhagenism. He soon made these doubts more concrete by actually constructing a model of the world in which the electron is an ordinary object and which agrees with the predictions of quantum theory.

Bohm's model manages to evade the terms of von Neumann's proof because although Bohm's electrons are objects—they have attributes of their own—the electron's attributes behave in a most peculiar manner. Suppose the cards in quantum monte actually were quantum objects in the manner of Bohm's electrons. Three cards are laid out and they are *all black*. You choose your first card and as you do, the other cards both change their colors to red! Now when you turn over the next card, you will not get a pair. The act of turning over one card caused the other cards to change their colors!

In Bohm's model each electron is sensitively attuned to everything that's going on in its environment, especially the presence of a measuring device. As you set up your experiment, the electron "changes its colors" depending on what you decide to measure. You register the value not just of some passive attribute but of an attribute that is determined in part by your own actions. The electron senses what's going on around it via a new kind of field called the "pilot wave." Bohm's pilot wave puts the electron in instant contact with every other particle in the universe. The instantaneousness of this connection prevents you from trying to "trick the electron" by making two simultaneous measurements at distant locations—in effect turning over both cards at the same time. The electron's pilot wave moves too fast for such tricks to work.

Since Bohm's model predicts the same results as quantum theory and yet is made of little objects, von Neumann's proof obviously contains a loophole: his notion of "ordinary objects" is too restrictive. Copenhagenists protest that entities as outlandish as Bohm's hardly deserve to be called "objects." An entity that can instantly change its properties in response to a tiny change made half a universe away is no ordinary object. Orders for such a change would have to move faster than light, which Einstein has shown to be impossible. Although Bohm's electrons—"little superluminal chameleons"—are a kind of object that might conceivably underlie the quantum facts, Copenhagenists reject their existence on other grounds—namely, 1. "objects" should not be in touch with everything in the universe, and 2. especially not faster than light.

Bohm found he could explain the quantum facts with an underlying reality consisting of objects, but the bizarre properties of such objects dash any hope that Bohm's solution represents a return to ordinary reality. Bohm's success in circumventing von Neumann's proof encouraged others to construct different neorealist models of quantum reality, but all of these models also contained Bohm's "defect"—this embarrassing instant connection of each object to every other. CERN physicist John Bell, as part of his celebrated theorem, showed that these instant connections are no accident, but a necessary feature of any object-based model of reality and of many other models of reality as well.

Quantum Reality #7: (Consciousness creates reality.) The first person to suggest that quantum theory implies that reality is created by human consciousness was not some crank on the fringes of physics but the eminent mathematician John von Neumann. In his quantum bible *Die Grundlagen,* the most influential book on quantum theory ever written, von Neumann concludes that, from a strictly logical point of view, only the presence of consciousness can solve the measurement problem. As a professional mathematician, von Neumann was accustomed to boldly following a logical argument wherever it might lead. Here, however, was a severe test for his professionalism, for his logic leads to a particularly unpalatable conclusion: that the world is not objectively real but depends on the mind of the observer.

Von Neumann's argument may be difficult to believe but it is easy to understand. He begins by assuming that quantum theory is correct and applies to everything in the world, measuring devices in particular. This means that everything in the world is represented by a proxy wave. Bohr, you will recall, granted M devices a special exemption from the quantum description; in the Copenhagen interpretation, M devices enjoy a special classical-style status. On the other hand, in von Neumann's all-quantum model measurement devices are not special but are represented by proxy waves like everything else.

The price von Neumann must pay for treating all entities alike is the need for a wave function collapse. Somewhere between the system being observed and the mind of the observer, the proxy wave, instead of *expanding* to fill all its possibilities (the natural tendency of proxy waves everywhere), must *contract* into just one possibility—the actual measurement result. What's special in von Neumann's model is not the measuring device but the *measurement act*, where many waves suddenly shrink into one. Von Neumann showed that this special act—the wave function collapse or "quantum jump"—could be located anywhere without changing the final results, but it could not be eliminated. In von Neumann's all-quantum description of the world, the quantum measurement problem boils down to one question: where in fact does the wave function collapse?

Between the measured system itself and the observer's mind stretches a series of intermediate devices (each represented by its own proxy wave) called the von Neumann chain. Solving the measurement problem on von Neumann's terms means finding the location at which nature breaks this chain with a quantum jump.

Von Neumann's version of the measurement problem would be easily resolved if we could just regard the wave function collapse not as an actual physical process but as a mere artifact of the theory—a spurious detail present in the math but not in the real world. When dice come up seven, for instance, we do not regard this as a "collapse" of spread-out dice possibility into one particular actuality. Why then should we regard the quantum probability collapse as any more real than the realization of a dice event?

The trouble with this easy solution is that the quantum ignorance which lies behind quantum probability is different in kind from the classical ignorance responsible for dice odds. According to the orthodox ontology, every individual quon is exactly alike until the quantum jump occurs: hence this jump is more than an individual realization of laws that hold for ensembles; the quantum jump is a sudden change in the rules that influence single events.

Every link in von Neumann's chain (with one exception) consists of aggregates of quons interacting with other quons. Since all quons are fundamentally the same, these interactions are all qualitatively alike. Outside a measurement situation, a quon's wave function has no tendency to collapse. How can a collection of quons possibly know whether it's located "in the wild" or inside some M device? In a system consisting solely of quons, no special site seems to exist which might host a quantum leap; there's no natural boundary line between measurer and measuree.

Some physicists have proposed that the essence of measurement is *making a record*. Since making a record always involves *an irreversible event*—else the record would fade—they suggest that von Neumann's chain breaks naturally at the first thermodynamically irreversible process. However, despite much talk about how thermal disorder *might* hold the key to the measurement problem, the fact remains that no one has succeeded in showing how random activity by itself suffices to turn quantum ignorance (where the desired information simply doesn't exist) into classical ignorance (where the desired information exists but is hidden).

The existence of measurements in which "nothing happens" (Renninger-style measurement), where knowledge is gained by *the absence of a detection,* is also difficult to reconcile with the view that irreversible acts cause quantum jumps. In a Renninger-style measurement, there must always be the "possibility of an irreversible act" (a detector must actually be present in the null channel), but this detector does not click during the actual measurement. If we take seriously the notion that irreversible acts collapse the wave function, Renninger measurements require us to believe that the mere possibility of an irreversible act is sufficient to bring about a quantum jump. The fact that such "interactionless" measurements are possible means that the wave function collapse cannot be identified with some specific random process occurring inside a measuring device.

The only unusual link in the von Neumann chain is the link between the observer's brain and the observer's mind. Here, where the mystery of matter gives way to the mystery of mind, is a privileged position for the elusive quantum jump.

The gist of von Neumann's argument is this: in order for an all-quantum world to work, a special process—the quantum jump—must be present in all measurement acts and nowhere else. But in a world made entirely of quantumstuff there are no privileged processes. The only special process that lies outside matter's monopoly is the awareness of the observer.

Here's how physicists Fritz London and Edmond Bauer sum up the case for Quantum Reality #7:

"A coupling even with a measuring device is not yet a measurement. A measurement is achieved only when the position of the pointer has been *observed*. It is precisely this increase of knowledge acquired by observation that gives the observer the right to choose among the different components of the mixture predicted by theory, to reject those which are not observed, and to attribute henceforth to the object a new wave function, that of the pure case which he has found. We note the essential role played by the consciousness of the observer in this transition from the mixture to the pure case. Without his effective intervention, one would never obtain a new ψ function . . . Thus it is not a mysterious interaction between the apparatus and the [quantum] object that produces a new ψ for the system during measurement. It is only the consciousness of an 'I' who can, by virtue of his observation, *set up a new objectivity.*"

Consciousness-created reality (QR #7) should not be confused with mere observer-created reality (QR #2). Quantum realists belonging to these schools make very different claims. Any observer—conscious or not—has to make a choice of what attributes to measure (quantum meter option), which determines into which waveforms the quantum system will be analyzed. By his choice of what to measure, the observer will cause the quon to take on position rather than momentum attributes, but he does not decide what the value of this quon's position shall be. The quantum meter option can just as well be exercised by an inanimate computer as by a human observer. The observer "creates reality" here by choosing *what kinds of attributes a quon shall possess.* (Observer creation of the first kind.)

Consciousness-created reality goes one step farther. Consciousness selects (or at least acts as the site for such a selection) which one of the many position possibilities actually becomes realized. Thus the meter option selects what game shall be played (position instead of momentum, for instance); consciousness actually deals out the cards (this particular value of position). Consciousness "creates reality" by deciding *what particular attribute value shall materialize.* (Observer creation of the second kind.)

According to Quantum Reality #7, dynamic attributes, when not being observed, exist as a wavewise superposition of possibilities; the universe acquires definite values for these attributes only during a conscious observation. A mere machine can't manufacture reality, in this view, unless it embodies some kind of awareness analogous to our own; the measurement problem is solved by a ghost in the machine. This quantum reality suggests that most of the universe most of the time dwells in a half-real limbo of possibility waiting for a conscious observer to make it fully real.

During the eighteenth century the growing success of Newton's clockwork mechanics inclined many philosophers to the belief that *all phenomena*, including life, mind, and spirit could ultimately be explained as types of complex machinery. George Berkeley, bishop of Cloyne in southern Ireland, outraged by scientific materialism, opposed it with strong philosophical opinions of his own. Berkeley argued that mind is not a form of matter but quite the opposite: matter does not even exist except as the perception of some mind. Absolute existence belongs to minds alone—the mind of God, the minds of humans and other spiritual beings. All other forms of being, including matter, light, the Earth, and stars, exist only by virtue of some mind's being aware of them. In Berkeley's philosophy— dubbed "idealism" because it emphasizes the primacy of ideas over things —nothing exists unless it is either a mind itself, or is perceived by a mind. *Esse est percipi* (To be is to be perceived) was the Irish bishop's motto concerning matter: "All those bodies which compose the mighty frame of the world have no subsistence without a mind."

Quantum idealism does not go as far as Berkeley's. According to Quantum Reality #7, all quons and their static attributes enjoy an absolute existence whether they are observed or not. Only a quon's dynamic attributes, including the major external attributes position and momentum, are mind-created. Thus all those entities "which compose the mighty frame of the world" do certainly *exist* without the intervention of mind, but until someone actually looks at them, these entities possess no definite place or motion. The peculiar existential status of *unobserved* quons is the topic of my eighth quantum reality.

Quantum Reality #8: The duplex world of Werner Heisenberg. No matter how bizarre the visions of quantum realities dancing in their heads, most physicists agree that *the results of measurements are truly real.* Like ordinary people (but unlike some philosophers), physicists cannot deny the evidence of their senses. The indubitable reality of measurement results is a solid rock on which to found an empirical science, or from which to launch speculative voyages into deep reality.

In most quantum realities the measurement act does not passively reveal some preexisting attributes of quantum entities, but actively transforms "what's really there" into some form compatible with ordinary experience. One of the main quantum facts of life is that we radically change whatever we observe. Legendary King Midas never knew the feel of silk or a human hand after everything he touched turned to gold. Humans are stuck in a similar Midas-like predicament: we can't directly experience the true texture of reality because *everything we touch turns to matter.*

Many of the previous quantum realities have focused on what extra feature makes an ordinary interaction into a measurement (macroscopic device, record-making observer, conscious spectator, for example), but little has been said about the character of the unmeasured state. Since most of reality most of the time dwells in this unmeasured condition, which quantum theory represents by an uncollapsed superposition of possibilities, the lack of such a description leaves the majority of the universe (everything that's not currently being measured) shrouded in mystery.

Werner Heisenberg was one of the few physicists who attempted to describe in non-mathematical terms the "world-in-itself," that innocent existence quantum entities enjoy before undergoing a measurement. Heisenberg's description is no full-fledged model of reality, but just one man's attempt to convey in ordinary language the flavor of the deep reality symbolized by a ψ wave.

Heisenberg was acutely aware of the difficulty of trying to describe the quantum world in words. "The problems of language here are really serious," he admits. "We wish to speak in some way about the structure of the atoms and not only about the 'facts'—for instance, the water droplets in a cloud chamber. But we cannot speak about the atoms in ordinary language." Heisenberg however did not let this difficulty render him speechless. He realized that some words are better than others for describing the unmeasured world.

Quantum theory according to the Copenhagen interpretation represents the world in two different ways: the observer's experience is expressed in the classical language of *actualities,* while the unmeasured quantum realm is represented as a wavewise superposition of *possibilities.* Heisenberg suggests that we take these representations literally as a model for the way things really are.

Thus, according to Heisenberg's duplex vision, the unmeasured world *actually is* what quantum theory represents it to be: a superposition of mere possibilities (Heisenberg called them *potentia)*, unrealized tendencies for action, awaiting the magic moment of measurement that will grant one of these tendencies a more concrete style of being which we humans experience as actuality.

Just as traffic noise contains no tubas or pianos but might be wave-analyzed in terms of tuba or piano waveforms, so also is it a mistake to imagine Heisenberg's world of potentia as consisting of definite preexisting possibilities for certain courses of action. Superposition is not like that. The possibilities latent in a particular proxy wave emerge only in a definite measurement context. Not until the observer exercises his quantum meter option can he speak of a superposition of definite possibilities. Heisenberg's world of potentia is not only empty of actualities, even its *possibilities* are not as well defined, in the absence of a measurement situation, as, say, classical dice possibilities.

Heisenberg's potentia represents a novel kind of physical existence standing "halfway between the idea of the event and the actual event itself." Until it's actually observed, a quantum entity must be considered "less real" than the same entity observed. On the other hand, an unobserved quantum entity possesses "more reality" than that available to ordinary objects because it can entertain *in potentia* a multitude of contradictory attributes which would be impossible for any fully actualized entity. "In the experiments about atomic events we have to do with things and facts, with phenomena that are just as real as any phenomena in daily life," says Heisenberg. "But the atoms or the elementary particles themselves are not as real; they form a world of potentialities or possibilities rather than one of things or facts."

Heisenberg's half-real universe of potentia is reminiscent of certain oriental views developed in contexts far removed from quantum physics:

> This floating world is but a phantasm
> It is a momentary smoke

Though ghostly and transitory, Heisenberg's shimmering ocean of potentia is the sole support for everything we see around us. The entire visible universe, what Bishop Berkeley called "the mighty frame of the world," rests ultimately on a strange quantum kind of being no more substantial than a promise.

QUANTUM REALITY REPRISE

Most physicists use quantum theory as mere recipe for calculating results and don't trouble themselves about "reality." However, it is hard to believe that this theory could be so successful without corresponding in some way to the way things really are. Like the story of the blind man and the elephant in which each man imagines a different animal according to which part he's holding, these eight quantum realities result from different physicists each taking a part of quantum theory seriously and identifying it with the "real reality" behind appearances.

Bohr, for instance, took the *uncertainty principle* seriously, using it to argue that quons possess no dynamic attributes of their own. A quon's so-called attributes reside in the *relation* between the entity itself and a "classical" measuring device.

Believers in observer-created reality take the *quantum meter option* seriously: exercising this powerful option, the observer decides which kinds of attributes quantum entities will seem to possess.

Quantum Reality #3 (undivided wholeness) takes *phase entanglement* seriously as a token of a real physical connectedness (the quantum connection) that instantly joins each quon to every other.

Everett in his many-worlds interpretation takes the *quantum measurement problem* seriously: he solves this problem by declaring that the wave function collapse is an illusion caused by human inability to experience reality fully: we are unaccountably blind to all but a single branch of a luxuriant tree of coexistent universes.

Quantum logicians (QR #5) view *incompatible attributes* as the central quantum fact: a new form of reasoning is needed in which the strange behavior of quantum attributes appears perfectly natural.

Neorealists (QR #6) believe that the *surface ordinariness of all quantum facts* (Cinderella effect)—the fact that every experiment must be described in classical language—clearly suggests that reality itself is ordinary too.

Believers in consciousness-created reality take the *quantum measurement problem* seriously and conclude that nowhere in mere quantumstuff is there a logical site for proxy wave collapse: only the mind of the observer can fill the bill.

Werner Heisenberg's duplex universe (QR #8) takes literally the *proxy wave representation of quantum entities* as superpositions of possibilities. Unmeasured quantum attributes are just what quantum theory says they are: unrealized possibilities.

Because of the Cinderella effect—the stubborn ordinariness of quantum fact—we cannot experience directly any of these strange quantum realities (with the exception of neorealism). Although these realities make very different claims, they all predict exactly the same facts. At present there's no way to decide experimentally among these alternative visions of the way the world really is.

When my son asks me what the world is made of, I confidently answer that deep down, matter is made of atoms. However, when he asks me what atoms are like, I cannot answer though I have spent half my life exploring this question. How dishonest I feel—as "expert" in atomic reality—whenever I draw for schoolchildren the popular planetary picture of the atom; it was known to be a lie even in their grandparents' day.

Physicists cannot explain atoms to their children, not because we are ignorant but because we know too much. The behavior of atoms is no longer a mystery. Quantum theorists can confidently calculate the outcome of any conceivable atomic experiment. However, as we see, the price physicists have paid for quantum theory's remarkable predictive power is their inability to picture in plain language an image of the atomic world.

Thirty years after the publication of von Neumann's *Die Grundlagen* (1932) the quantum reality question inspired heated debates among philosophers and physicists, but little progress was made toward solving the problem of what sort of world we actually live in. Then, in 1964, a physicist named John Bell proved an important theorem which gave us a new insight into deep reality.

11

The Einstein-Podolsky-Rosen Paradox

One can escape from this conclusion only by either assuming that the measurement of B (telepathically) changes the real situation at G or by denying independent real situations as such to things which are spatially separated from each other. Both alternatives appear to me entirely unacceptable.

Albert Einstein

The gist of Bell's theorem is this: *no local model of reality can explain the results of a particular experiment.* In short: reality is non-local. Before we examine what "non-locality" means, let's take a look at this particular experiment, called the EPR experiment, which is the factual basis for Bell's important result. Like so many other innovations in twentieth-century physics, the EPR experiment was conceived by Albert Einstein.

Although he helped put it together, Einstein was never satisfied with quantum theory. He didn't like its intrinsic randomness ("I cannot believe that God plays dice with the universe"), but most of all he disliked the fact that quantum theory (as interpreted by Bohr and Heisenberg) implies that reality is observer-created. "I cannot imagine," Einstein once said,

Albert Einstein

Boris Podolsky

Nathan Rosen

FIG. 11.1 *Albert Einstein, Boris Podolsky, and Nathan Rosen, originators of the EPR paradox which purports to demonstrate the existence of extra "elements of reality" not included in quantum theory.*

"that a mouse could drastically change the universe by merely looking at it." Einstein accused Bohr and Heisenberg of attempting to restore man (and mouse) to the center of the cosmos from which Copernicus had ousted them nearly five hundred years ago. "The belief in an external world independent of the perceiving subject," Einstein maintained, "is the basis of all natural science."

Bohr responded by comparing Einstein to the critics of his own relativity theory. He pointed out that thanks to Einstein's work, physicists have come to realize that space and time are not absolute but relative to an observer's state of motion. In quantum theory we simply take this way of thinking one step further and recognize that reality itself (or at least its dynamic attributes) is also observer-dependent. Why did Einstein find it so difficult, Bohr wondered, to accept this natural extension of his own ideas?

"A good joke should not be repeated twice," Einstein quipped.

Niels Bohr and Albert Einstein debated the quantum reality question for as long as they lived: Einstein failed in his attempts to assault quantum theory head on, and reluctantly agreed with Bohr that quantum theory describes correctly all presently conceivable experiments—a conclusion that remains uncontested today. Einstein resorted instead to criticizing quantum theory on the grounds that it is *incomplete*.

Quantum theory may be sufficient to explain experiments, Einstein confessed, but experiments are only part of what goes on in the world. Because quantum theory makes only statistical predictions, it cannot help but leave out certain "elements of reality" which a more adequate theory of the world must include.

Niels Bohr, on the other hand, claimed that although quantum theory does give only statistical predictions, it is still complete. Quantum theory's indefiniteness is a virtue, not a weakness, because it corresponds to an indefiniteness that actually exists in the world. It is foolish to seek a precise description of an imprecise world; such misplaced precision is bound to miss the mark.

Einstein put forth his best argument for quantum theory's incompleteness in the form of a thought experiment involving two correlated quons. He devised this experiment at Princeton in 1935 with the help of two American physicists: Boris Podolsky, originally from southern Russia, and Brooklyn-born Nathan Rosen. The original Einstein, Podolsky, and Rosen (EPR) experiment concerned two *momentum-correlated electrons*, but physicists today repeat EPR's argument using David Bohm's conceptually simpler experiment involving two *polarization-correlated photons*.

THE EPR EXPERIMENT

In Chapter 8, I compared a light beam to a series of balls (photons) thrown by a baseball pitcher. The two-valued photon polarization attribute was compared to a batter holding his bat at a certain angle ϕ and getting either a hit or a miss. In the laboratory, photon polarization is measured with a calcite crystal which splits a light beam into up and down channels depending on whether its photons are polarized along or across the calcite's optic axis.

The EPR experiment is only slightly more complicated than this two-man ball game. The EPR source emits *pairs of photons* (Green and Blue) which travel in opposite directions to *two distant detectors* (also labeled Green and Blue) where their polarization $P(\phi)$ at a particular angle ϕ can be measured. To visualize this EPR arrangement we imagine a pitcher who throws *two balls at a time*. First he throws a Green ball to home plate; then, without breaking rhythm, he turns and fires a Blue ball to second base where a second batter is waiting.

As in the previous game, the batters at home and at second can each measure the "polarization" of the baseball by holding their bats at a particular angle. A hit shows the ball to be polarized at the bat angle; a miss means polarization at right angles to the bat.

The pitcher fires off a pair of balls, rests for a moment, then throws another pair. For each pair of balls, the Green player measures his Green ball's polarization at some Green angle, while the Blue player measures her Blue ball's polarization at some Blue angle. To understand the EPR experiment, it's not necessary actually to know what polarization really is —what polarization "really is" is a mystery to physicists too—but only the particular results of each pair of polarization measurements. Encoded in the pattern of these results is the gist of the EPR paradox as well as the core of Bell's theorem, discussed in the next chapter.

The EPR photon pairs are pitched in a special way; they come out of the light source in a particular phase-entangled state called the "state of parallel polarization." Because their phases are entangled with each other, each photon's phase depends on what the other photon is doing. Consequently, neither photon by itself is represented by a definite waveform; hence (according to quantum theory) neither photon possesses a definite polarization.

Observationally, not possessing a definite polarization means that no measurement of polarization will always give the same result. In fact, for this particular two-photon state the Green light and the Blue light are *completely unpolarized*—the maximum indefiniteness possible for a two-valued attribute. For each photon at any angle ϕ, a polarization measurement $P(\phi)$ gives 50 percent up/50 percent down, results which occur at random, like flipping a coin.

Although *each photon by itself* does not possess a definite proxy wave, *the two-photon state as a whole* is represented by a definite wave, which means that certain *two-particle attributes* (which belong to the Green and Blue photon together) have a definite value. For photons in the state of parallel polarization, one such definite attribute is the photons' *paired polarization*.

To measure paired polarization $PP(\phi)$, at a particular angle ϕ, set both Green and Blue calcites at the same angle ϕ and look at their polarization values (up or down). Like polarization itself, the PP attribute can take two possible values: either both photons have the same P (match) or they have opposite P (miss).

Both quantum theory and quantum fact agree that for photons in the parallel polarization state, $PP(\phi)$ *at all angles* ϕ, always has the same value, namely match. This means that if you measure the Green polarization at angle ϕ and the Blue polarization at the same angle, both polarizations are always the same. Furthermore the P of G will be the same as the P of B no matter how far apart the photons fly or which polarization happens to be measured first. For instance, you can measure the polarization of the Green photon immediately after it leaves the source and measure the Blue photon a year later (when it is one light year away from its source): the polarizations of both photons will be identical.

According to quantum theory, in the state of parallel polarization each photon by itself has no definite P. However, the PP of G and B together is definite: it's match in every direction. The polarization attributes of unmeasured photons in this state resemble the attributes of identical twins before conception. Each twin's attributes (sex, hair color, and so forth) are undecided but the status of their *paired attributes* is already known: the same for both. For this reason I call the state of parallel polarization "the twin state."

In terms of the two baseball players, the results of a long series of plays against a pitcher who always throws pairs of balls in the twin state is this:

1. At no matter what angle ϕ either player holds the bat, he/she always gets a 50-50 mixture of hits and misses;

2. If both players agree beforehand to hold their bats at the same angle ϕ (I call this move "measuring the $PP(\phi)$ attribute), whatever happens to one player's ball (hit or miss) also happens to the other player's ball.

IS QUANTUM THEORY A COMPLETE DESCRIPTION OF REALITY?

One difference between human twins and a pair of photons in the twin state is that before conception the human twins are nonexistent, while before measurement the photons already exist. We know that they were emitted at a certain time from their source and are traveling with a certain velocity toward their respective detectors.

For a pair of photons in the twin state, Einstein asked the question, "Is the P of photon G, after it's emitted but before it's actually measured, *truly indefinite* as Bohr's interpretation of quantum theory requires, or is it, like identical twins in the womb, *really definite* but unknown?" In other words, "Is our uncertainty concerning the unobserved polarizations a matter of *quantum* or *classical* ignorance?"

According to Bohr, the P of photon G does not even exist before we measure it. G's so-called attributes belong not to the photon itself but reside partly in "the entire experimental arrangement." Like the position of a rainbow, polarization is a *relational attribute* and does not come into existence until Green observer decides how he will deploy his apparatus at location G (and possibly elsewhere as well). It's nonsense to suppose that before a measurement, photon G has some definite polarization. Einstein argues that, on the contrary, not only does photon G have a definite P in some direction, it has a definite P *in every direction*.

To dramatize the difference between Bohr and Einstein, let's imagine that Blue player moves closer to the mound so that she gets her Blue ball before Green player gets his. Suppose she holds her bat at zero degrees (vertically) and gets a hit, which means that her photon is V-polarized. We now switch our camera to home plate where the spirits of Bohr and Einstein are discussing the reality status of the as-yet-to-be measured Green photon presently hurtling toward the Green batter at the speed of light. To allow the great men time for debate, we imagine the usual passage of time to be temporarily suspended.

BOHR: When I say that quantum theory is "complete," I mean that QT says everything that can possibly be said about the reality of that Green photon. If it's not in the theory, it's not in the photon either.

EINSTEIN: What, then, does quantum theory say about this Green photon now approaching the Green batter?

BOHR: In the first place, given that Blue's already measured a V photon, coupled with the fact that this pitcher throws nothing but twin-state photon pairs, quantum theory predicts that if Green chooses to hold his bat vertically, he will certainly get a hit; furthermore it also predicts that if he holds his bat horizontally, he will certainly get a miss.

EINSTEIN: I agree with you concerning what quantum theory predicts if Green makes either a horizontal or a vertical polarization measurement. Now, what is supposed to happen if Green holds his bat at some other angle?

BOHR: For Green bat angles other than zero or ninety degrees, quantum theory gives no certain results, but only the relative *probability* of a hit. For instance, if Green should hold his bat at 45 degrees, the odds are 50-50 that he will get a hit.

EINSTEIN: Yes. Quantum theory indeed gives only statistical predictions for intermediate angles. We seem to agree concerning the predictions of the theory and about the facts of the matter—namely, that quantum theory has never made a single incorrect prediction. We agree, as Kant would have put it, about the appearances and about the theory. But what, my dear Bohr, are you willing to say about the *reality* of this particular Green photon magically suspended before us?

BOHR: Because I believe that quantum theory describes all physical situations completely, I must say that before it is actually measured, this photon really has a definite polarization only in the *V* and *H* directions, but no others. To speak of a definite polarization in any other directions would be to talk nonsense. Thus I say that, in reality, this Green photon does not possess polarization attributes except perhaps at these exceptional angles.

Even at these special angles, for which quantum theory gives certain results, I am not entirely convinced that these results represent a definite attribute belonging solely to the photon. I believe that all attributes are joint creations of photon and measuring device and do not belong to one or the other.

EINSTEIN: Concerning this matter of completeness . . . As you know, my friend, I cannot refute your opinion that quantum theory is *a complete theory of phenomena:* it indeed seems to describe correctly the results of every experiment my poor head has been able to imagine. But I do not share your faith that quantum theory is *a complete theory of reality.* I believe that certain elements of reality exist in the world that are not described by the quantum formalism. In the case of this Green photon, for example, I say that it possesses a definite polarization attribute for every possible angle, not just for the *V* and *H* directions.

BOHR: No, my friend, you are mistaken. Except perhaps in certain special situations where the outcome is not a matter of chance—such as the *V* and *H* directions in this case—the photon's polarization is a joint production of the entire experimental arrangement, and does not inhere in the photon by itself independent of a particular measurement context.

EINSTEIN: Forgive me, Bohr, but I have never been able to understand your subtle reasoning in this matter. Indeed, for situations like this twin-state baseball game, I have, with my colleagues Podolsky and Rosen, devised a simple argument which convinces us that this Green photon hovering in front of us possesses a definite (but unknown) polarization attribute *at every angle.* Permit me to show you this argument.

Our reasoning depends on a certain plausible assumption, which physicists nowadays call "the locality assumption": we assume that the real factual situation of the Green photon, after it's left the source, is not affected by how the Blue player chooses to hold her bat. In other words, we assume that *Blue's batting stance does not affect the Green photon.* This supposition seems reasonable since both photons are traveling in opposite directions at the speed of light. Therefore one photon cannot learn about the other's measurement situation except via signals that travel faster than light.

BOHR: I am suspicious of this locality assumption but please continue.

EINSTEIN: Here is our argument. For this present situation, Blue chose to hold her bat vertically and she got a hit. But if she had held her bat at some other angle, say 45 degrees, she would also have measured *something*, either a hit or a miss, we do not know which. Because this photon pair is in the twin state we know that Green photon would be obliged to show the same polarization that Blue got at 45 degrees. In like manner Blue could have held her bat at any angle X and measured a certain polarization; Green photon is compelled to have an identical polarization at angle X.

If Green photon must have a definite polarization for each Blue measurement choice, and if (by the locality assumption) Blue's measurement choice does not physically affect the Green photon, then *the Green photon must already possess a definite polarization for each angle* —polarizations that exist regardless of Blue's actual choice.

Thus we believe we have shown that before it strikes the Green bat, this Green photon has already "made up its mind" as to how it will act no matter how Green might choose to hold his bat. This Green photon must possess a sort of hit/miss list which tells it what to do for every bat angle. Quantum theory, on the other hand, certainly does not recognize any such list: except for the H and V directions, it considers these results to be "random," utterly unknown except in a probabilistic sense. Quantum theory is therefore "incomplete" because it leaves out some attributes—this hit/miss list, for example—which this photon seems to possess.

BOHR: Your argument is clever but I cannot accept your conclusion. Of course there is no question of any *mechanical influence* traveling from Blue's bat to the Green photon, but *there is essentially the question of an influence on the very conditions which define the possible types of predictions regarding the future behavior of the Green light.*

EINSTEIN: Yes, I remember your making that very statement in 1935 in response to our original EPR paper. I did not understand it then, and despite considerable effort, I must confess that I still cannot grasp the subtlety of your thought on this matter.

Since the author seems to have frozen our intellects, like that time-suspended Green photon out there, into our ancient philosophical positions, I will answer your old quote with two of my own which sum up my thinking on the EPR experiment:

"We are forced (via the EPR argument) to conclude that the quantum-mechanical description of physical reality given by wave functions is not complete."

"One can escape from this conclusion only by either assuming that the measurement of B (telepathically) changes the real situation at G, or by denying independent real situations as such to things which are spatially separated. Both alternatives appear to me entirely unacceptable."

Bohr, Einstein, and numerous other thinkers struggled with the EPR paradox but no generally acceptable solution could be found until Bell focused attention on the fragility of the locality assumption. Let's take a closer look at this locality assumption so essential to the argument of Einstein, Podolsky, and Rosen.

THE LOCALITY ASSUMPTION

The locality assumption does not mean that *what happens* at the Green bat has nothing to do with *what happens* at the Blue bat. Since the photons are correlated at the light source, the results at the Green and Blue measurement sites will likewise be correlated. What locality means is that no action on Blue's part (as she detects her Blue photon) can affect what Green player sees (when he detects his Green photon). Locality means that what happens at home plate is unaffected by how Blue holds her bat at second base.

The locality assumption is necessary to EPR's argument because although Blue observer could have made any polarization measurement she pleased, she can in fact (for a single photon) make only one, because photon polarizations at different angles are incompatible attributes.

As a homely example of EPR's reasoning, consider a shop (Enrico's Pizza Reale) which sells three different pizzas: Sicilian, Milanese, and Neapolitan. Whenever you order a pizza from Enrico's it arrives at your door in ten minutes. Since a pizza takes thirty minutes to bake, you know that the pizza you ordered must have been ready when you phoned.

Suppose you order a pizza of your choice each night (but you can only afford one), and it's always delivered in ten minutes. Can you conclude that Enrico keeps on hand *all three kinds of pizza?*

Not without a kind of locality assumption. You have to assume that Enrico has no way of knowing what kind of pizza you are going to order that night. If he can discover your choice beforehand, he need keep only one pizza hot.

Your nightly freedom of choice plus the (no pizza spies) "locality assumption" allows you to infer, on the basis of a series of one-pizza observations, that Enrico in reality keeps *all three pizzas* ready to go each night. The argument for preexisting polarizations is the same as for preexisting pizzas. Blue player's freedom to choose her single *P* measurement plus the locality assumption allows EPR to infer that *all polarizations* must be simultaneously present in the Green photon (in the form of a hit/miss list) before Green player makes his measurement.

Hence, in the twin state, photon *G* already secretly knows how it will respond to any polarization measurement that Green player might care to make upon it. According to EPR's argument, *Green photon's polarization attribute is not indefinite at all.* Green photon's hit/miss list specifies its polarization at all measurement angles.

Bohr asserts that photon *G*, before it's measured, is in an indefinite state of polarization: quantum theory does not recognize any such hit/miss list. But Einstein, Podolsky, and Rosen can *prove* that such a list exists in nature. Hence according to EPR, quantum theory is necessarily incomplete.

It is important to realize what EPR did not do: Einstein, Podolsky, and Rosen did not find an experimental situation where quantum theory is factually wrong. What EPR discovered was a simple logical argument (based on the experimental fact of perfect polarization correlation in a certain two-photon system) that *indirectly demonstrates* the existence of photon attributes which quantum theory fails to take into account. EPR then ask, "If quantum theory is a complete theory of reality, why does it omit these attributes?"

What's at stake here is not whether quantum theory is a complete theory of *phenomena* (accounting correctly for all presently conceivable measurements) but whether it is a complete theory of *reality* (accounting correctly for whatever exists whether measurable or not). Many "refutations" of the EPR argument consist merely of demonstrating that quantum theory describes correctly all twin-state polarization measurements. EPR do not contest quantum theory's competence to describe phenomena; Einstein, Podolsky, and Rosen claim, however, to have demonstrated the existence of certain "elements of reality" (in Einstein's words), parts of the world *not directly observable* which quantum theory simply leaves out.

The EPR proof gives those who believe that what's real is only what can be observed an opportunity to put their convictions to the test. For such no-nonsense realists, the argument of EPR which purports to demonstrate the existence of an extra-observational reality must be mistaken. However, even those convinced beforehand of EPR's error found it surprisingly difficult to point out the fallacy in their reasoning. Hundreds of papers were published on the "EPR paradox." For thirty years physicists and philosophers beat their heads against this proof without either refuting EPR's logic or shedding further light on EPR's alleged "elements of reality."

In 1964 the long-standing EPR stalemate was broken by the efforts of theorist John Bell.

12

Bell's Interconnectedness Theorem

Contagious magic is based upon the assumption that substances which were once joined together possess a continuing linkage; thus an act carried out upon a smaller unit will affect the larger unit even though they are physically separated.

Sir James Frazer

John Stewart Bell was born and grew up in Belfast, Northern Ireland. He is now a theoretical physicist at CERN (a large accelerator center in Geneva financed by Western European countries) where he specializes in elementary particle physics. In 1964, while on sabbatical leave from CERN, Bell decided to investigate the quantum reality question, which had fascinated him since his undergraduate days.

Bell began by looking at von Neumann's proof, which demonstrates the impossibility of neorealism. According to von Neumann, the world cannot be made of ordinary objects, which possess dynamic attributes of their own. Bell discovered that although this proof excludes objects whose attributes combine in "reasonable ways," it does not forbid objects which can change their attributes in response to their environment. This loophole in von Neumann's proof is what allows Bohm, de Broglie, and other neorealists to build explicit ordinary-object-based models of quantum reality: all these models contain objects whose attributes are context-sensitive.

While preparing a review article on von Neumann's proof, Bell became interested in impossibility proofs in general and wondered whether a proof could be constructed which would conclusively exclude any model of reality that possessed certain physical characteristics. Bell himself managed to devise such a proof which rejects all models of reality possessing the property of "locality." This proof has since become known as *Bell's theorem*. It asserts that no local model of reality can underlie the quantum facts. Bell's theorem says that reality must be non-local.

In a letter to the author, John Bell recalls his discovery: "I had for long been fascinated by EPR. Was there a paradox or not? I was deeply impressed by Einstein's reservations about quantum mechanics and his views of it as an incomplete theory. For several reasons the time was ripe for me to tackle the problem head on. The result was the reverse of what I had hoped. But I was delighted—in a region of wooliness and obscurity to have come upon something hard and clear."

The structure of Bell's proof is as follows. For a certain class of two-quon experiments (the EPR experiment and its variations), Bell *assumes* that a local reality exists. With a bit of arithmetic he shows that this locality assumption leads directly to a certain inequality (Bell's inequality) which the experimental results must satisfy. Whenever these experiments are done, they violate Bell's inequality. Hence the local-reality assumption is mistaken. Conclusion: any reality that underlies the EPR experiment must be non-local.

WHAT IS A LOCAL INTERACTION?

The essence of a local interaction is direct contact—as basic as a punch in the nose. Body A affects body B *locally* when it either touches B or touches something else that touches B. A gear train is a typical local mechanism. Motion passes from one gear wheel to another in an unbroken chain. Break the chain by taking out a single gear and the movement cannot continue. Without something there to mediate it, a local interaction cannot cross a gap.

On the other hand, the essence of non-locality is unmediated action-at-a-distance. A non-local interaction jumps from body A to body B without touching anything in between. Voodoo injury is an example of a non-local interaction. When a voodoo practitioner sticks a pin in her doll, the distant target is (supposedly) instantly wounded, although nothing actually travels from doll to victim. Believers in voodoo claim that an action *here* causes an effect *there;* that's all there is to it. Without benefit of mediation, a non-local interaction effortlessly flashes across the void.

The unruly nature of unmediated action has moved physicists from Galileo to Gell-Mann to unanimously reject non-local interactions as a basis for explaining what goes on in the world. No one has so vehemently expressed physicists' distaste for non-local interactions as Sir Isaac Newton:

"That one body may act upon another at a distance through a vacuum without the mediation of anything else . . . is to me so great an absurdity, that I believe no man, who has in philosophical matters a competent faculty for thinking, can ever fall into."

Given his antipathy for non-local forces, Newton was somewhat embarrassed by his own theory of gravity. If a non-local force is "so great an absurdity," how does the sun's gravity manage to cross millions of miles of empty space to hold the Earth in its orbit? Concerning the actual nature of gravity, Newton wisely held his tongue. *"Hypotheses non fingo,"* he declared. "I frame no hypotheses."

Newton's faith in strictly local forces was vindicated by his successors, who explained gravity in terms of the *field concept.* The space between the sun and Earth is not empty, today's physicists say: it's filled with a gravitational field which exerts a force on any body it touches. The modern field concept allows us to regard gravity as a strictly local interaction even though it acts across vast reaches of space. The sun's mass produces a gravity field; this field pulls on the Earth and mediates the sun-Earth interaction.

Physicists today share Newton's belief that the world is tied together by strictly local connections. All presently known interactions can be explained in terms of only four fundamental forces (strong, weak, electromagnetic, and gravitational). In every case these forces act as if they are mediated by fields. Since quantum theory has blurred the once sharp distinction between particle and field (both are quantumstuff now) we can equally well say these local forces are mediated by the exchange of *particles.* Thus the sun attracts the Earth (and vice versa) via the gravity field or via an exchange of gravitons (the particle aspect of the gravity field). In actuality gravity (as is true for the other three fundamental forces as well) is carried neither by particle or field but by something that partakes of both, an innately quantum go-between whose mediation makes every one of nature's forces strictly local.

Although the concept of locality does not strictly demand it, most forces diminish in strength as you move away from their source. It is conceivable that a local force might stay constant or even increase with distance from its source (the force of a stretched spring, for instance, increases with distance). The big four forces that hold the world together happen, however, all to *decrease with distance*—gravity and electromagnetism diminish as the inverse square; the strong and weak forces fall off considerably faster.

The toughest limitation on a local interaction is how fast it can travel. When you move an object A, you stretch its attached field. This field distorts first near object A, then the field warp moves off to distant regions. Einstein's special theory of relativity restricts the velocity of this field deformation to light speed or below. According to Einstein, no material object can travel faster than light; not even the less material field warp can travel so fast.

Non-local influences, if they existed, would not be mediated by fields or by anything else. When A connects to B non-locally, nothing crosses the intervening space, hence no amount of interposed matter can shield this interaction.

Non-local influences do not diminish with distance. They are as potent at a million miles as at a millimeter.

Non-local influences act instantaneously. The speed of their transmission is not limited by the velocity of light.

A non-local interaction links up one location with another without crossing space, without decay, and without delay. A non-local interaction is, in short, *unmediated, unmitigated,* and *immediate.*

Despite physicists' traditional rejection of non-local interactions, despite the fact that all known forces are incontestably local, despite Einstein's prohibition against superluminal connections, and despite the fact that no experiment has ever shown a single case of unmediated faster-than-light communication, Bell maintains that the world is filled with innumerable non-local influences. Furthermore these unmediated connections are present not only in rare and exotic circumstances, but underlie all the events of everyday life. Non-local connections are ubiquitous because reality itself is non-local.

Not all physicists believe Bell's proof to be an airtight demonstration of the necessary existence of non-local connections. But the alternatives these critics offer instead seem to me to be generally obscure and/or preposterous. As we shall see in the following chapter, some physicists will go so far as to actually "deny reality itself" rather than accept Bell's audacious conclusion that quantum reality must be non-local.

How Bell Proved Reality Cannot Be Local

To understand the import of Bell's theorem and the arguments of his critics, it's necessary to look at Bell's proof in some detail. Fortunately Bell's theorem is easier to prove than the Pythagorean theorem taught in every high school. The simplicity of Bell's proof opens it to everyone, not just physicists and mathematicians.

Bell's proof is based on the same EPR experiment used by Einstein, Podolsky, and Rosen to demonstrate the existence of hidden "elements of reality" which quantum theory neglects to describe. The "EPR paradox" consists of the fact that for thirty years physicists have neither been able to refute EPR's argument nor shed further light on EPR's alleged "elements of reality."

The EPR experiment involves a source of light which produces pairs of photons (Green and Blue) in the "twin state." These photons travel in opposite directions to calcite detectors (G and B) which can measure their polarization attribute $P(\phi)$ at some angle ϕ. In the twin state each beam by itself appears completely unpolarized—an unpredictably random 50-50 mixture of ups and downs at whatever angle you choose to measure.

Though separately unpolarized, each photon's polarization is perfectly correlated with its partner's. If you measure the P of both photons at the same angle (a two-photon attribute I call paired polarization), these polarizations always match.

For his proof, Bell considers another two-photon attribute called polarization correlation *(PC)* which can be measured on these photons. Attribute *PC* is measured the same way as attribute *PP* except that the calcites are set not at the same but at different angles. To measure *PC(θ)*, set Green calcite at a particular angle ϕ_G and Blue calcite at angle ϕ_B. Now compare Green and Blue polarizations for each pair of photons. If these

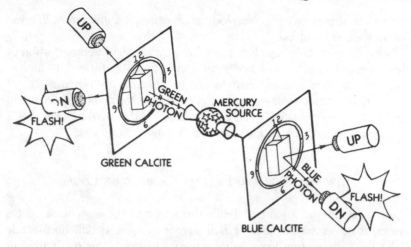

FIG. 12.1 *The EPR experiment. The central mercury source emits pairs of pho-tons (Green and Blue) in the twin state. At Green and Blue measuring sites, the polarization P(0) of each of these photons is recorded with a calcite-based P meter. Bell's theorem concerns the unusual strength of the polarization correlation existing between these Green and Blue photons.*

Ps are the same (both up or both down), call this a match; if opposites, call this a miss. Angle θ is the angle between the two calcites, namely $\theta = \phi_G - \phi_B$.

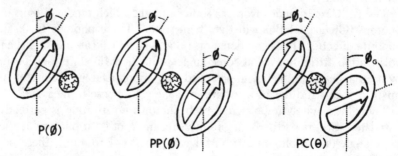

$P(\phi)$ $PP(\phi)$ $PC(\theta)$

FIG. 12.2 *Three kinds of polarization measurement. A. Measuring P(ϕ)—ordinary polarization—involves counting the number of photons polarized along (up) or across (down) the calcite's optic axis oriented at angle ϕ. B. Measuring PP(ϕ)—paired polarization—involves comparing the polarization of two photons at the same angle ϕ (miss or match). C. Measuring PC(θ)—polarization correlation—involves comparing the polarization of two photons at two different angles (θ is the angular difference between the two calcite settings).*

For photons in the twin state, quantum theory predicts that PC $(\phi_G - \phi_B)$ depends only on the *relative angle* θ between calcites and is independent of the separate settings ϕ_G and ϕ_B. Thus if the angle of the Green calcite differs by 30° (in either direction) from that of the Blue calcite, the value of $PC(30)$ will be the same, no matter how Green and Blue happen to be tilted. The fact that $PC(\theta)$ depends only on the *difference* between the two calcites has been amply verified by experiment.

For each angle θ between calcites, a PC measurement asks for the *fraction of matches* obtained in a long series of photon pairs. Thus $PC = 1$ means all matches (no misses) while $PC = 0$ means no matches (all misses). Bell's theorem is concerned with how this match fraction changes as we vary the angle between calcites from zero to ninety degrees.

For our previous discussion of the twin state, we already know the value of PC at zero and ninety degrees. At a calcite separation of zero degrees, $PC = 1$. When both calcites are set at the same angle, a PC measurement is identical to what I've called a PP measurement, which for the twin state yields a 100 percent match at all angles.

At a calcite separation of ninety degrees, $PC = 0$. When a calcite is turned through a right angle, its photon-sorting operation is merely reversed: its up channel passes downs and vice versa. At ninety degrees a P meter behaves like the same P meter at zero degrees with its outputs relabeled. This calcite channel reversal plus the perfect correlation at zero degrees leads to a *perfect anti-correlation* when the calcite axes differ by ninety degrees.

At zero degrees, $PC = 1$; at ninety degrees, $PC = 0$. In between, PC varies between 1 and 0 as the angle between calcites swings from zero to ninety degrees. The meat of Bell's proof is *the actual shape of this variation.*

To dramatize what's happening in this EPR experiment, imagine that Green detector is on Earth, and Blue detector is on Betelgeuse (540 light-years away) while twin-state correlated light is coming from a spaceship parked halfway in between. Although in its laboratory versions the EPR experiment spans only a room-size distance, the immense dimensions of this thought experiment remind us that, in principle, photon correlations don't depend on distance.

The spaceship acts as a kind of interstellar lighthouse directing a Green light beam to earth, a Blue light beam to Betelgeuse in the opposite direction. Forget for the moment that Green and Blue detectors are measuring something called "polarization" and regard their outputs as coded messages from the spaceship. Two synchronized binary message sequences composed of ups and downs emerge from calcite crystals 500 light-years apart. How these two messages are connected is the concern of Bell's proof.

When both calcites are set at the same angle (say, twelve o'clock), then $PC = 1$. Green polarization matches perfectly with Blue. Two typical synchronized sequences of distant P measurements might look like this:

GREEN: U D U D D U D D D U U D U D D U
BLUE: U D U D D U D D D U U D U D D U

If we construe these polarization measurements as binary message sequences, then whenever the calcites are lined up, the Blue observer on Betelgeuse gets the same message as the Green observer on Earth.

Since PC varies from 1 to 0 as we change the relative calcite angle, there will be some angle α at which $PC = 3/4$. At this angle, for every *four* photon pairs, the number of matches (on the average) is *three* while the number of misses is *one*. At this particular calcite separation, a sequence of P measurements might look like this:

*   * *   *

GREEN: U D D D D U D D D U D D U D D U
BLUE: U D U D D D U D D U U D U D D U

At angle α, the messages received by Green and Blue are not the same but contain "errors"—G's message differs from B's message by one miss in every four marks.

Now we are ready to demonstrate Bell's proof. Watch closely; this proof is so short that it goes by fast. Align the calcites at twelve o'clock. Observe that the messages are identical. Move the Green calcite by α degrees. Note that the messages are no longer the same but contain "errors"—one miss out of every four marks. Move the Green calcite back to twelve and these errors disappear; the messages are the same again. Whenever Green moves his calcite by α degrees in either direction, we see the messages differ by one character out of four. Moving the Green calcite back to twelve noon restores the identity of the two messages.

The same thing happens on Betelgeuse. With both calcites set at twelve noon, messages are identical. When Blue moves her calcite by α degrees in either direction, we see the messages differ by one part in four. Moving the Blue calcite back to twelve noon restores the identity of the two messages.

Everything described so far concerns the results of certain correlation experiments which can be verified in the laboratory. Now we make an assumption about what might actually be going on—a supposition which cannot be directly verified: the locality assumption, which is the core of Bell's proof.

We assume that *turning the Blue calcite can change only the Blue message;* likewise turning the Green calcite can change only the Green message. This is Bell's famous locality assumption. It is identical to the assumption Einstein made in his EPR paradox: that Blue observer's acts cannot affect Green observer's results. The locality assumption—that Blue's acts don't change Green's code—seems entirely reasonable: how could an action on Betelgeuse change what's happening right now on Earth? However, as we shall see, this "reasonable" assumption leads immediately to an experimental prediction which is contrary to fact. Let's see what this locality assumption forces us to conclude about the outcome of possible experiments.

With both calcites originally set at twelve noon, turn Blue calcite by α degrees, and at the same time turn Green calcite *in the opposite direction* by α degrees. Now the calcites are misaligned by 2α degrees. What is the new error rate?

Since turning Blue calcite α degrees puts one miss in the Blue sequence (for every four marks) and turning the Green calcite α degrees puts one miss in the Green sequence, we might naïvely guess that when we turn both calcites we will get exactly two misses per four marks. However, this guess ignores the possibility that a "Blue error" might fall on the same mark as a "Green error"—a coincidence which produces an apparent match and restores character identity. Taking into account the possibility of such "error-correcting overlaps," we revise our error estimate and predict that whenever the calcites are misaligned by 2α degrees, the error rate will be two misses—*or less.*

This prediction is an example of a *Bell inequality.* This Bell inequality says: If the error rate at angle α is 1/4, then the error rate at twice this angle cannot be greater than 2/4.

This Bell inequality follows from the locality assumption and makes a definite prediction concerning the value of the PC attribute at a certain angle for photon pairs in the twin state. It predicts that when the calcites are misaligned by 2α degrees the difference between the Green and Blue polarization sequences will not exceed two misses out of four marks. The quantum facts, however, say otherwise. John Clauser and Stuart Freedman carried out this EPR experiment at Berkeley and showed that a calcite separation of 2α degrees gives three misses for every four marks—a quite substantial violation of the Bell inequality.

Clauser's experiment conclusively violates the Bell inequality. Hence one of the assumptions that went into its derivation must be false. But Bell's argument uses mainly facts that can be verified—photon PCs at particular angles. The only assumption not experimentally accessible is the locality assumption. Since it leads to a prediction that strongly disagrees with experimental results, this locality assumption must be wrong. To save the appearances, we must deny locality.

Denying locality means accepting the conclusion that when Blue observer turns her calcite on Betelgeuse she instantly changes some of Green's code on Earth. In other words, locations B and G some five hundred light years apart are linked somehow by a non-local interaction. This experimental refutation of the locality assumption is the factual basis of Bell's theorem: no local reality can underlie the results of the EPR experiment.

Einstein, Podolsky, and Rosen used the locality assumption to demonstrate the existence of hidden "elements of reality" which quantum theory fails to take into account. However, if Blue and Green observers are linked by a non-local interaction, as the factual violation of the Bell inequality seems to imply, then EPR's argument is invalid by virtue of a false premise. The failure of their argument does not prove, of course, that no such "elements of reality" exist, but only that one cannot make a case for their existence by using EPR's reasoning. The logical necessity of non-local interactions resolves the EPR paradox (in Bell's words) "in the way which Einstein would have liked the least."

Reviewing the EPR paradox in his autobiography, Einstein reaffirmed his faith in locality: "On one supposition we should, in my opinion, absolutely hold fast: the real factual situation of the system (G) is independent of what is done with the system (B) which is spatially separated from the former." Einstein did not live to see Bell's proof and would certainly have been surprised by Bell's refutation of his cherished postulate. But I think he would have welcomed the strange news Bell's theorem brings us concerning the true nature of the quantum world. Bell's result vindicates Einstein's intuition that something funny is going on in quantum-correlated two-particle states.

As in the case of the EPR paradox, it's important to realize what Bell did not do. He did not discover an experimental situation in which non-

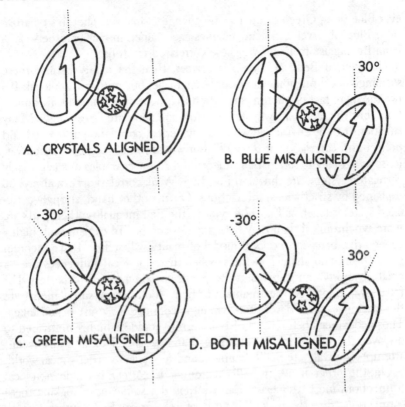

FIG. 12.3 *Simple proof of Bell's theorem A. Both calcites aligned: errors = zero. B. Blue crystal tilted by 30°: errors = 1 in four marks. C. Green crystal tilted by −30°: errors = 1 in four marks. D. Both Green and Blue crystals tilted by 30°; what's the new error?*

If we assume locality (Green's move cannot change Blue's mark), this new error cannot be greater than 2 in four marks. However, for EPR photons the PC measurement at 60° gives 3 errors in four marks. Therefore locality assumption is false.

local interactions are directly observed. Instead he invented a simple argument based on experimental results that *indirectly demonstrates* the necessary existence of non-local connections.

The phenomena displayed by photon pairs in the twin state are entirely *local*. The only spin-space attribute accessible to Green observer is his Green photon polarization $P(\phi)$. This attribute is always 50–50 random (unpolarized) no matter how Blue observer sets her calcite. Because whatever Blue does, Green can detect no change in his own photon's polarization, Blue observer can send no message—superluminal or otherwise—from Betelgeuse to Earth via these correlated photons.

However, if Bell's argument is correct, then the reality behind these seemingly local phenomena not only might be, but *must be* non-local. It's not the mere fact of photon correlation that necessitates non-local connections, but the fact that twin-state photons are correlated *so strongly*. Many situations can be envisioned which show perfect correlation at $\theta = 0°$ and perfect anti-correlation at $\theta = 90°$, but whose in-between correlation varies so as actually to *satisfy* Bell's inequality. A few examples of such weakly correlated systems are shown in Fig. 12.5. Weak correlations can always be explained by strictly local interactions. On the other hand, strongly-correlated systems (such as Fig. 12.4) violate the Bell inequality; their parts are more synchronized than they have any right to be. To explain such highly cooperative behavior, no local model of reality will suffice. Bell's theorem gives those who share Newton's belief that non-local influences are "a great absurdity" an opportunity to put their convictions to the test. For folks loyal to locality, the argument of Bell which purports to demonstrate the existence of hidden faster-than-light connections must be mistaken. Those convinced beforehand of Bell's error should be highly motivated to discover the fallacy in his reasoning. Later we will look at some recent attempts to invalidate Bell's argument and to recover a strictly local world.

On the other hand, if Bell's reasoning is correct invisible non-local connections must truly exist. Can we then devise means of making these connections directly evident instead of relying on Bell's indirect argument? The possibility of practical superluminal communication via the quantum connection will be discussed in the next chapter.

Bell proved his theorem for a particular two-photon system. What justification exists for extending his conclusion (the reality underlying the EPR experiment must be non-local) to the general case of everyday experience (the reality underlying *everything* must be non-local)? To expand the scope of Bell's argument we turn to quantum theory.

In quantum theory's formalism, what accounts for strong photon correlation in the twin state is *phase entanglement*. Whenever quantum system A meets quantum system B, their phases get mixed up. Part of A's proxy wave goes off with B's wave and vice versa. Phase entanglement thereafter instantly connects any two quons which have once interacted. Before Bell's discovery, this strong quantum connection had been recognized (especially by Schrödinger, who considered it quantum theory's most dis-

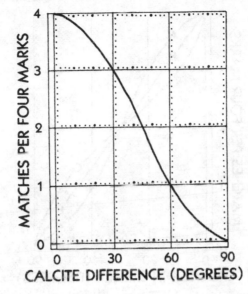

FIG. 12.4   *Quantum theory's prediction for PC (θ) of twin-state mercury light. This result violates Bell's inequality, hence argues against locality. In 1972 this prediction became a matter of fact (measured by Clauser at Berkeley); now the quantum facts say locality is false.*

tinctive feature) but regarded by physicists as a kind of mathematical fiction with no roots in reality. Since Bell's theorem demands a superluminal connection and quantum theory provides one—in the form of ubiquitous but presumably "fictitious" phase connections—perhaps these quantum connections are not as fictitious as was once believed.

Since there is nothing that is not ultimately a quantum system, if the quantum phase connection is "real," then it links *all systems that have once interacted at some time in the past*—not just twin-state photons—into a single waveform whose remotest parts are joined in a manner unmediated, unmitigated, and immediate. The mechanism for this instant connectedness is not some invisible field that stretches from one part to the next, but the fact that a bit of each part's "being" is lodged in the other. Each quon leaves some of its "phase" in the other's care, and this phase exchange connects them forever after. What phase entanglement really is we may never know, but Bell's theorem tells us that it is no limp mathematical fiction but a reality to be reckoned with.

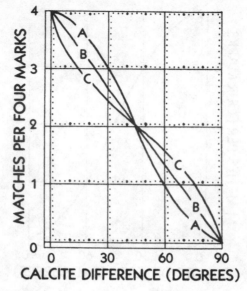

FIG. 12.5 *Some polarization correlations which do not violate Bell's inequality. Curve A is the strongly correlated mercury light pictured in Fig. 12.4. The straight line B satisfies Bell's inequality; so does the dished-in curve C. Both curves B and C can be simulated by local mechanisms; they are weakly correlated. Curve A cannot be simulated by any local mechanism. These curves illustrate on just how small a quantitative difference Bell's important theorem rests.*

CLAUSER'S EXPERIMENT

In 1964, when Bell derived his inequality, no twin-state PC measurements existed against which it could be tested. However, the *calculation* of twin-state polarization is an elementary exercise in quantum theory. This calculation predicts that $PC(\theta) = cos^2\theta$, a correlation plotted as Fig. 12.4. The angle α at which misses = 1/4 for $cos^2\theta$ is 30°. Bell's inequality consequently demands that the number of misses at 2α (60° in this case) shall be no greater than 2/4. However, at 60° this expression gives a miss fraction of 3/4. Since 3/4 is considerably greater than 2/4, the theoretical expression $PC = cos^2\theta$ violates Bell's inequality. This violation marks the twin state as a *strongly* correlated system—a pair of entities linked tighter than any local reality can explain.

The fact that this calculated result violates Bell's inequality implies that *any system which obeys these quantum-theoretical predictions cannot be explained by a local reality.* Before Bell's discovery, one could still imagine that a local reality lurked beneath the experimental facts; after 1964, one could blissfully believe in a strictly local world only by hoping that *quantum theory was wrong* in its predictions concerning photons in the twin state.

Since it challenges one of physicists' most cherished beliefs—that the world is fundamentally local—one might have expected Bell's proof to explode like a bombshell in the corridors of science. Instead, Bell's proof, published in an obscure little journal, was largely ignored even by those physicists who managed to find out about it.

Most physicists are not impressed by Bell's proof because it deals with *reality,* not *phenomena.* The majority of physicists are phenomenalists—whose professional world is circumscribed by phenomena and mathematics. A phenomenalist perceives science as advancing in two directions: 1. new experiments uncover novel phenomena; 2. new mathematics explain or predict phenomena in original ways. Since it proposes no new experiments and derives no new phenomena-relevant mathematics, but merely puts certain constraints on an invisible reality, Bell's proof lies outside the fashionable formula for success in science and is generally dismissed by scientists as "mere philosophy."

Physicists' cool reception of Bell's proof is reminiscent of David Hume's famous prescription for separating truth from nonsense: "Does it contain any abstract reasoning concerning quantity or number? *No.* Does it contain any experimental reasoning concerning matter of fact and existence? *No.* Commit it then to the flames: for it can contain nothing but sophistry and illusion."

In the midst of this climate of indifference toward theories of reality, John Clauser, a young Ph.D. from Columbia, proposed actually to measure twin-state photons to see whether their polarization correlation attribute satisfied Bell's inequality (world is local; quantum theory wrong) or not (world is non-local; quantum theory right). Clauser received no support at Columbia for his proposal to put Bell's inequality to experimental test, and moved to Berkeley where apparatus already existed which he knew he could modify to measure twin-state photons.

Many kinds of excited atoms emit pairs of twin-state photons as they return to their ground state. Most experiments carried out to test Bell's inequality have used either mercury atoms excited by electron impact or calcium atoms excited by laser light. Clauser's Berkeley mercury source operates like a mercury-vapor streetlamp—both emit Blue and Green twin-state light—but Clauser's source was smaller and more intense than the lamps which nightly flood Telegraph Avenue with photons in the twin state.

Because real photon detectors are not 100 percent efficient—they count only about 10 percent of the photons which strike their phosphor faceplates—one cannot simply compare Bell's inequality to experimental results. Adapting Bell's original reasoning to existing experimental realities, Clauser and his colleagues derived a version of Bell's inequality (called the "CHSH inequality" after John F. Clauser, Michael A. Horne, Abner Shimony, and Richard A. Holt) which is testable with low-efficiency detectors.

Clauser was motivated to test the Bell inequality by his strong faith that the world was ultimately local. If quantum theory predicts a result which conflicts (via Bell's proof) with locality, so much the worse for quantum theory. Clauser anticipated that his experiment would prove quantum theory wrong at least in this matter of twin-state polarization. The results proved otherwise. In 1972 Clauser announced that quantum theory had passed his test. Bell's inequality had been experimentally violated by Blue and Green photons at Berkeley. Now not merely quantum theory but quantum fact contradict the hypothesis that the world is linked up by strictly local lines.

ASPECT'S EXPERIMENT

Clauser's pioneer test of the Bell inequality contains a loophole through which a desperate logician might still derive a local world. To pinpoint this loophole, let's return to our imaginary EPR experiment in space.

Clauser's mercury source sent Blue and Green light to opposite corners of a room. Our spaceship lighthouse shoots photons to Betelgeuse and Earth five hundred light-years apart.

Clauser switched the orientation of his P meters every 100 seconds. Clauser's switching time, translated to cosmic lighthouse scale, corresponds to keeping the P meters on Earth and Betelgeuse fixed *for more than a billion years.* Such leisurely P measurements would permit information on how distant P meters were set to leak between Earth and Betelgeuse at sublight speed (carried perhaps in the gossip of interstellar tourists)—information which could allow most of the photons to simulate strong correlations by strictly local means. To block the possibility of sub-luminal security leaks during long P-meter rests, the experimenter must be able to change the P meters while the photons are in flight. To change a calcite this fast in the lab means switching its orientation in a few bil-lionths of a second.

Unfortunately, mere matter just can't move that fast. However, physi-cist Alain Aspect at the University of Paris devised an experiment to test Bell's inequality which uses two acousto-optical switches to deflect each photon beam to one or the other of two preset calcite detectors. Instead of rapidly moving his calcites, Aspect moves his light beams.

With his ultrafast switches, Aspect can measure a different polarization every 10 billionth of a second, fast enough to eliminate subluminal secu-rity leaks between Blue and Green P meters. If Aspect's twin-state pho-tons violate Bell's inequality, the reality that underlies their strong correla-tion must connect Green and Blue measurement stations at a speed exceeding the velocity of light. Aspect completed his experiment in 1982, verifying the strongly correlated quantum theoretical predictions, hence violating Bell's inequality and supporting his contention that our phenom-enally local world is in actuality supported by an invisible reality which is unmediated, unmitigated, and faster than light.

Although Bell's theorem arose in the context of quantum theory, Bell's result does not depend on the truth of quantum theory. The Clauser-Aspect experiments show that Bell's inequality is violated by the facts. This means that even if quantum theory should someday fail, its successor theory must likewise violate Bell's inequality when it comes to explaining the twin state. Physics theories are not eternal. When quantum theory joins the ranks of phlogiston, caloric, and the luminiferous ether in the physics junkyard, Bell's theorem will still be valid. Because it's based on facts, Bell's theorem is here to stay.

IMPOSSIBLE WORLDS

Bell's theorem is an important tool for reality research because it enables folks who create imaginary worlds confidently to reject millions of impossi-ble worlds at a single glance. Bell's theorem tells you right away: "If it's local, it's hokum."

One of the worlds soundly obliterated by Bell's proof is the "distur-

bance model" of quantum reality. In this model—a species of neorealism
—quantum entities actually possess attributes of their own whether measured or not, but the measuring device changes these attributes in an
unpredictable and uncontrollable way. The inevitable disturbance of the
quantum system by the device which measures it gives rise, in this model
of reality, to quantum randomness, to the uncertainty principle and all the
other quantum oddities.

As a picture of how the quantum world might actually operate, many
physicists who have not given much thought to the matter take refuge in
some vague disturbance model of reality. For several years I avoided thinking about the quantum reality question by supposing that a disturbance
model of some kind was sufficient to account for the strange quantum
facts.

Such a disturbance model would explain, for instance, the observed
polarization of the Green photon in the EPR experiment as a result of the
Green calcite's "uncontrollable disturbance" of some intrinsic Green photon attribute. In other words, this model explains Green observer's results
by appealing to a hypothetical mechanism which involves only the Green
photon and the Green calcite. Bell's theorem shows that any such local
mechanism, no matter how ingenious, simply fails to fit the quantum
facts: Bell's proof knocks out the disturbance model because it's local.

Facile popular expositions often invoke the disturbance model of measurement to justify Heisenberg's uncertainty principle: we cannot know a
quantum entity as it is because we must inevitably disturb whatever we
observe. Bell's result shows this notion of quantum measurement as local
disturbance to be as outdated as the obsolete picture of the atom as
miniature solar system.

Another type of impossible world is the "classical style" reality symbolized by Newton's apple. Apples, and everything else in such a world, are
truly ordinary objects which possess attributes all their own even when not
being measured. When measured, whether by man, woman, or machine, a
classical apple merely reveals some attributes which it previously possessed.

Such an apple world (which experts call a "local non-contextual reality") is not inconceivable or illogical. But, according to Bell's theorem,
apple world is impossible because it can't possibly fit the facts. As a model
for the world we actually live in, apple world and all its local non-contextual cousins are, by virtue of their locality, sheer fantasy worlds.

We obviously need to be more sophisticated in our choice of possible worlds. Let's imagine, for instance, a *relational reality* patterned after the notions of Niels Bohr. The entities that make up such a world are like rainbows: they do not possess definite attributes except under definite measurement conditions. Upon measurement, attributes do emerge but they are a joint possession of entity and M device. In such a rainbow reality (called "local contextual"), attributes are not innate to an entity but change when the conditions of observation change. The only restriction we place upon such observer-induced changes is that distant M devices cannot change an entity's condition if such an influence would require a faster-than-light signal. In such a contextual, but local, reality, only nearby observers take part in the determination of an entity's apparent attributes.

Like apple world, rainbow world is neither inconceivable nor illogical. It is simply, on account of its locality, not the sort of world we happen to live in.

Bell's theorem rejects apple worlds; it also rejects rainbow worlds. What kinds of worlds does Bell's theorem allow?

A Possible World

Imagine Joe Green, an inhabitant of a *non-local* contextual world. Up in his sky, Joe sees a rainbow made up of a glistening pattern of colored dots. Unlike the regular dots in a photographic halftone, Joe's rainbow's dots form a random array.

On the other side of the same sun lies a counter-Earth, where Suzie Blue watches another rainbow in her counter-sky. Suzie's rainbow is likewise composed of a random array of colored dots. When Joe Green moves his chair, his rainbow moves too (a rainbow's position attribute is contextual, not innate), but Suzie's rainbow stands still. However, when Joe moves his chair Suzie's random array 200 million miles away instantly changes into a different (but equally random) array of colored dots. Suzie is not aware of this change—one random array looks pretty much like any other—but this change actually happens whether she notices it or not.

The *phenomenon* in this hypothetical world, whether the rainbow moves or not, is completely local: Suzie's rainbow doesn't move when Joe changes places. However, this world's *reality*—the array of little dots that make up both rainbows—is non-local: Suzie's dots change instantly whenever Joe moves his chair.

Such a non-local contextual world, in which stable rainbows are woven upon a faster-than-light fabric, is an example of the kind of world permitted by Bell's theorem. A universe that displays *local phenomena* built upon a *non-local reality* is the only sort of world consistent with known facts and Bell's proof. Superluminal rainbow world could be the kind of world we live in.

During the past twenty years Bell's theorem has been proved in many ways, some of which refer to photon attributes and some which don't. My version of Bell's proof makes no essential use of the concept of a photon or its attributes. Although Green and Blue photons and their polarization attributes are mentioned to familiarize you with the details of the EPR experiment, when it comes to the proof of Bell's theorem my argument is formulated entirely in terms of a pair of binary messages printed by particular macroscopic objects. I prove Bell's theorem here in terms of moves (orientations of calcite crystals) and marks (ups and downs on a data tape).

Bell's theorem as a relation between moves and marks takes non-locality out of the inaccessible microworld and situates it squarely in the familiar world of cats and bathtubs. Expressed in thoroughly macroscopic language, Bell's theorem says: *In reality, Green's move must change Blue's mark non-locally.* From arguments based on phenomena alone (no appeal to hidden attributes) we conclude that clicks in a certain counter must be instantly connected to the movement of a distant crystal of calcite.

For anyone interested in reality, Bell's theorem is a remarkable intellectual achievement. Starting with fact plus a bit of arithmetic, Bell goes beyond the facts to describe the contours of reality itself. Although no one has ever seen or suspected a single non-local phenomenon, Bell proves conclusively that the world behind phenomena must be non-local.

If all the world's phenomena are strictly local, what need is there to support local phenomena with a non-local fabric? Here we confront an alien design sense bizarre by human standards: the world seems strangely overbuilt. In addition the world's superluminal underpinning is almost completely concealed—non-locality would have been discovered long ago if it were more evident; it leaves its mark only indirectly through the impossibly strong correlations of certain obscure quantum systems.

In his celebrated theorem, Bell does not merely suggest or hint that reality is non-local, he actually proves it, invoking the clarity and power of mathematical reasoning. This compulsory feature of Bell's proof particularly irks physicists whose taste in realities is strictly local.

Bell's important proof has caused a furor in reality research comparable

John Stewart Bell

FIG. 12.6 *CERN physicist John Stewart Bell, inventor of the interconnectedness theorem, which establishes non-locality as a general feature of this world.*

to the Einstein-Podolsky-Rosen scandal of 1935. On the one hand, Bell's theorem proves the existence of an invisible non-local reality. Those who prefer their realities to be local have so far not been able to refute Bell's argument. The fact that Bell's proof is remarkably clear and brief has not hastened its refutation.

On the other hand, although Bell's theorem *indirectly* necessitates a deep non-locality, no one has come up with a way to *directly* display this purported non-locality, such as a faster-than-light communication scheme based on these deep quantum connections. If reality research's bottom line is "Reality has consequences," then this Bell-mandated deep reality has so far failed to make a showing. What the future holds for Bell's instantly connected but as yet inaccessible deep reality is anyone's guess.

13

The Future
of Quantum Reality

Strikes at the source of daily confusion
Eliminates ignorant fantasy
Scrubbing your senses free of Illusion
Enjoy the taste of Reality;
Scrubbing your senses free of Illusion
Enjoy the taste of Reality.

> "Reality Calypso" from
> *Benjamin Bunny Faces Reality*

Bell's proof has the logical form *reductio ad absurdum*, namely:

1. Make an assumption A;
2. Show that it leads to a contradiction; hence
3. Conclude that assumption A is false.

To prove his theorem, Bell assumes a local reality, shows (with a bit of arithmetic) that this contradicts experimental fact, then concludes that no local reality can underlie these facts. Hence reality is non-local.

Reductio ad absurdum proofs are particularly vulnerable to refutation because it's impossible to make just one assumption. Along with major assumption A, lots of little minor assumptions a, b, c, d, inevitably creep in. If you're particularly fond of assumption A, you can derail the proof of its absurdity by blaming the logical contradiction on assumption b instead.

In the case of Bell's proof, one can continue to believe in a local reality by denying one of Bell's other assumptions instead. However, because Bell's proof is so short and most of his assumptions accessible to experiment, such additional suppositions are not easy to find. Hence the various negations of Bell's conclusion tend to be rather farfetched and lead to realities more preposterous than the superluminal reality they attempt to exorcise.

ATTEMPTS TO REFUTE BELL'S PROOF

Consider, for example, Cornell physicist N. D. Mermin's curious argument that since Bell assumes a local reality, we can blame the contradiction either on the locality assumption or on the assumption of "reality." Mermin wants to keep locality, so as far as he is concerned, reality must go. Mermin uses Bell's *reductio ad absurdum* method to conclude that "the moon is demonstrably not there when nobody looks."

It's difficult to convey to outsiders the distaste which the majority of physicists feel when they hear the word "non-locality." Physics, after all, is full of odd notions. What's so repulsive about a mere faster-than-light connection? Mermin's argument illustrates, more than words can say, the deep antipathy most physicists bear toward unmediated action-at-a-distance: these guys so treasure locality that they are willing to deny reality itself before accepting a world that's non-local.

However, when you hear a physicist saying that he "denies reality," take it with a grain of salt. Don't imagine the poor guy believes that everything's a dream. He just means to say that the reality he's got in mind is a bit unconventional. Physicists who deny reality altogether cease to be physicists and their insights do not appear in physics journals.

The form of Mermin's conclusion shows that he believes in a reality of the observer-created variety—what I've called "rainbow reality" because of the observer-dependence of the rainbow's position attribute. Rainbow reality can come in two flavors: *local rainbow reality* (Green observer creates only Green code) or *non-local rainbow reality* (Green observer helps create Blue code too). It's easy to show (via Bell's argument) that a local observer-created reality contradicts the quantum facts just as surely as does Bell's original local (but otherwise unspecified) reality.

Mermin's desperate attempt to escape the demon of non-locality by accepting a reality that's observer-created simply doesn't work. The mild kind of observer-created realism sanctioned by quantum theory—namely, that dynamic attributes are *relational*—has to be non-local to fit the facts. Opt for a rainbow reality if you will, but it's got to be superluminal.

Many scientists read the weekly journal *Science* to learn what's happening outside their fields. Reporting on the Aspect experiment, *Science* staff writer Arthur Robinson speaks for many physicists when he observes that Bell's assumption A is actually *two assumptions:* locality plus realism. "Realism," according to Robinson, "requires that the [polarization attributes] exist and have definite values whether or not they are measured." Instead of giving up locality, suggests Robinson, why not simply abandon realism? However, those who have followed me so far will see that Robinson's realism corresponds to what I call "apple world": a universe in which entities possess their attributes in the familiar classical manner independent of observation. Even the color of beefsteak is not "real" according to Robinson's standards.

Banishing Bell-guaranteed non-locality by giving up apple world would be a cheap victory indeed. If it would restore locality, most physicists would gladly relinquish classical-style reality, especially when they recall that in 1932 John von Neumann had already proved that kind of reality to be incompatible with quantum theory. Unfortunately you can give up a lot more than apple reality and still be stuck with non-locality.

For instance you can deny the existence of photons and all their attributes, both static and dynamic (no physicist goes this far)—but the world that's left (mostly macroscopic measuring devices and their responses) still needs to be superluminally linked. Look at the version of Bell's theorem presented in the previous chapter: it assumes *nothing* concerning the hypothetical attributes of photons, but deals solely with the measurement results (marks) of measuring devices in specified configurations (moves). Provided that when you deny reality you retain a belief in the real existence of measuring devices, their settings, and their results, Bell's proof shows that the operations of such machines must be non-locally connected. Bell's quantum connection not only conjoins the attributes of invisible microentities, but links as well the actions of heavy apparatus made of steel, glass, and calcite. In the EPR experiment, Bell's deep non-locality leaks into the macroscopic world, manifesting as superstrong relations between distant detection devices.

In my opinion, you can escape Bell's conclusion (reality is non-local) by denying reality only if you go all the way and claim that macroscopic objects (including measuring devices) are somehow not really there. Bishop Berkeley did not believe in the existence of mountains, apples, or polarization meters, but few physicists—even those committed to locality at all costs—are willing to go so far. Denying the evidence of your senses seems a high price to pay just to win an argument. Perhaps the world *really is* non-local.

There are, however, more subtle objections to Bell's proof. To visualize what's called the "CFD assumption," let's return to the EPR experiment in the form of a spaceship lighthouse beaming Green photons to Earth and Blue to Betelgeuse. Consider the locality assumption in the form: Green's move does not change Blue's mark.

This form of the locality assumption means that whether Green sets his calcite at twelve o'clock or two has no effect on Blue's result. But for each particular photon pair, Green can set his calcite *at only one angle*. Because of this experimental fact of life, the locality assumption actually breaks up into two parts: 1. if Green *had set his calcite* at some angle other than the one he actually chose, definite (but unknown) results *would have occurred* for both Green and Blue; 2. Blue's result for this hypothetical setting would have been the same as Blue's result for Green's actual setting.

Part one of the locality assumption supposes that if Green had chosen to carry out some other P measurement, Blue would have obtained *some result*. This assumption (known to quantum reality aficionados as contrafactual definiteness, or CFD, for short) is untestable because for each photon pair, Green can make only one kind of P measurement. Einstein, Podolsky, and Rosen also had to assume CFD in their celebrated proof for the existence of extra "elements of reality" not described by quantum theory. In my pizza pie analogy, the CFD assumption means that I take for granted the notion that ordering any kind of pizza other than the one I did in fact order would have resulted in its delivery. This CFD assumption, that hypothetical actions would have led to definite outcomes, seems reasonable but it is by its very nature untestable since each event happens only once. You can order only one pizza this Saturday night; Green can only align his P meter in one direction for photon #1136.

Part two of the locality assumption brings in the notion of locality per se: it asserts that Green's hypothetical acts do not affect Blue's hypothetical results. Because locality is expressed in terms of hypothetical results—the results of choices we could have made but did not—we see that unless we assume CFD, we cannot even formulate the notion of locality in Bell's sense. Die-hard fans of a local universe seize on this logical loophole and attempt to refute Bell's conclusion by *denying CFD.*

One of the most prominent physicists trying to rescue locality by dumping CFD is John Wheeler. Wheeler denies CFD this way: "[Bell's theorem] deals with worlds that never were and never can be. The real world is what we care about here." Wheeler attempts to invalidate Bell's theorem not by denying reality, but by upholding it in a very strict sense.

Contrafactual reasoning is the basis for personal, business, and military planning. The outcome of chess games and more serious conflicts is controlled not so much by actual events as by hypothetical threats and possibilities that never happen, but which could have happened had you acted otherwise. Reasoning in terms of hypothetical outcomes takes concrete form in contingency plans for nuclear attack and in computer chess decision trees which simply take the CFD assumption for granted. Why CFD is justified in the case of the hypothetical outcomes of conceivable (but perhaps "unthinkable") nuclear options, but not in the case of equally macroscopic P meter outcomes is hard for me to understand.

Contrafactual reasoning is so deeply rooted in human thinking that it's difficult even for determined opponents of CFD to eliminate it from their arguments. For instance, in order for John Wheeler's celebrated delayed-choice experiment (discussed in Chapter 9) to make sense, he must compare the outcome of experiment A not with the outcome of experiment B but with *the outcome that would have occurred had experiment B been performed instead of A,* a manifest example of CFD-infected reasoning.

Another way of gauging the plausibility of the no-CFD objection to Bell's theorem is to ask, "In what sorts of conceivable worlds would CFD be a patently invalid assumption?" One such no-CFD world is a universe where *only one history* was ever possible in the first place. To speak of hypothetical results in such a one-track world would be to talk nonsense. This kind of no-CFD world is strictly deterministic in the Newtonian clockwork style. Because of the absence of real choices, Bell's theorem can not even be formulated in a strictly Newtonian universe.

One way to evade Bell's theorem's non-local consequences would be to devise a local model of the quantum world in which CFD is a patently invalid assumption. No one, to my knowledge, has been clever enough to come up with such a local no-CFD picture of the world compatible with quantum theory.

Other attempts to invalidate Bell's theorem involve challenging Western (i.e., Boolean) logic, denying Green observer's free will or practicing subtle variations on the no-CFD theme. The arguments against Bell's theorem (and their counterarguments) have become so recondite that a meeting of physicists on this topic sounds much like a congress of medieval theologians.

Physicists continue to debate whether Bell's theorem is airtight or not. However, the real question is not whether Bell can prove beyond doubt that reality is non-local, but whether the world is *in fact* non-local.

Do Non-local Connections Permit Superluminal Signaling?

Whether deep reality is truly non-local or not could be settled in an instant by the discovery of a single superluminal signal. If the world is in truth bound together everywhere by faster-than-light connections, can we exploit these links to send superfast messages to our friends? Such an accomplishment would not only directly validate Bell's conclusion, it would initiate a new era for humankind, making us masters of space and time.

If we could exploit the quantum connection, Superman flying faster than light, hence going backward in time to save Lois Lane, need not happen only in the movies. According to Einstein's relativity, superluminal signals would open up similar channels from the present to the past—channels that would allow people today to change what by conventional reckoning has already happened. The fact that FTL signaling entails backward causality is regarded by some physicists as a powerful argument against the possibility of such signaling (and against non-locality in general) but the achievement of an actual FTL transmitter would obviously invalidate such philosophical objections.

In the EPR photon lighthouse, the natural quantum process that blocks FTL signaling is quantum randomness. Put yourself on Betelgeuse with Blue observer. No matter how she sets her Blue crystal, she receives a message from the central spaceship which consists of a 50-50 random pattern of ups and downs. When Green observer on Earth moves his calcite we know (via Bell's theorem) that his actions must change Blue's sequence of marks. Some of her ups change to downs and vice versa; if this did not happen, the correlation would be *weaker* than is in fact observed. However, these changes in the details of Blue's marks involve a shift from one random pattern to another equally random pattern. Since all random sequences look alike (although there are many kinds of order, there is only one kind of randomness), Blue is not aware of this Green-initiated change. The situation seems to be that *Green can send superluminal messages but Blue cannot decode them.*

Even if we believe (with the support of Bell's theorem) in universal superluminal links, we must face the possibility that *such links are private lines* accessible to the workings of nature alone, and are blocked to human use by an undecipherable scrambler built of perfect quantum randomness.

EBERHARD'S PROOF CONTRA SUPERLUMINAL SIGNALING

To the welter of proofs discussed so far—von Neumann's proof (against an ordinary-reality explanation of quantum theory), EPR's proof (for extra "elements of reality" ignored by quantum theory), and Bell's proof (against a local deep reality)—I am compelled to add one more. If quantum theory is correct, then it is possible to prove that quantum measurements cannot be used to send signals faster than light.

This result, first obtained by Berkeley physicist Philippe Eberhard, generalizes Green's failure to send a recognizable signal to Blue via strongly correlated twin-state photons. Eberhard's proof states that even in the midst of a superluminal reality, *no quantum measurement results can be connected faster than light.*

Eberhard's proof, unlike Bell's, depends on the validity of quantum theory. Eberhard uses quantum theory to calculate the effect of one "quantum measurement" (such as Green's choice of move) on another "quantum measurement" (such as Blue's pattern of marks) carried out on a phase-entangled system such as a pair of photons in the twin state. Eberhard's calculation shows that whatever the behavior of individual marks may be (because it's strictly statistical, quantum theory does not concern itself with individual events), *the pattern of marks* does not depend at all on these faraway moves. Individual message bits may conceivably change faster than light; since these bits occur at random we can neither verify nor disprove this claim. But according to Eberhard, the pattern of such bits remains precisely the same no matter how the distant detector is manipulated.

A "quantum measurement" is defined as a *statistically discernible* difference. Since there is no statistical difference between random sequences, the two random Blue sequences that may result when Green makes two different moves do not count as two quantum measurements. Thus Eberhard's proof permits nature to send perfectly encrypted messages along FTL channels but denies humans access to such channels so long as their actions are bound by the rules of quantum theory.

Several ingenious schemes have been devised to evade Eberhard's proof, based primarily on attempting to exploit the ambiguity that exists over what actually constitutes a quantum measurement, but all such schemes have so far failed. Skeptical scientists compare the attempt to construct real superluminal communicators based on strong quantum correlations in the face of Eberhard's impossibility proof with attempts which flourished in the last century to devise perpetual motion machines in the face of the law of energy conservation.

Barring an unforeseen breakthrough in superluminal communication research, the small part of the physics community concerned with reality research now splits into two parts: those who can prove non-local influences but cannot exploit them, and those who don't believe in Bell-guaranteed FTL links but can't refute them. How this tension at the heart of physics will ultimately be resolved lies in the unknown future. Meanwhile let's look at how Bell's discovery has affected physicists' reality crisis. How does Bell's theorem change our perception of the eight quantum realities?

BELL'S THEOREM AND REALITY

Basically Bell changed our view of reality by raising the issue of non-locality. After Bell, any serious model of reality has to be either manifestly non-local or custom-designed to render the locality/non-locality distinction meaningless. In any case, in this post-Bell era every quantum realist must deal with non-locality in one way or another.

Quantum Reality #1: The Copenhagen interpretation, Part I (There is no deep reality). Bohr believed that quantum entities possess no dynamic attributes of their own: such attributes as we measure them to possess are a joint product of quon and M device. According to the Copenhagen interpretation a quon's so-called attributes (excluding its static attributes) belong not to the quon itself but to "the entire measurement situation."

Bell's theorem tells us that Bohr's insight that dynamic attributes are creatures of the measurement situation is essentially correct: the notion that a quon's attributes are innate must be abandoned because of its blatant locality. However, "the entire measurement situation" which determines a quon's attributes is more extensive than Bohr could have foreseen. Like everyone else in his day, Bohr was a firm believer in locality. However Bell's theorem demonstrates that "the entire measurement situation" which decides what values Blue's attributes will display must include the setting of Green's crystal located possibly half a galaxy away. Bohr built better than he knew: what happens in his "rainbow reality"—a world of relational attributes—must depend not only on observers nearby but superluminally on "the entire experimental arrangement" no matter how distant.

Quantum Reality #2: The world is created in the act of observation. The notion that quons acquire their dynamic attributes via the act of observation is a cornerstone of the Copenhagen interpretation. Bell's theorem merely expands the notion of observer to include the action of people and apparatus at arbitrarily distant locations—locations outside the reach of conventional light-speed-limited signals.

It's important to understand that Bell's theorem requires *reality*, not *phenomena*, to be superluminally linked. Measurement is the means by which a physicist makes contact with reality. All quantum measurements are made up of quantum jumps—the individual ups and downs of a particular M device, or the flashes on a phosphor screen, for example. Quantum phenomena consist of *the persistent and repeatable patterns* which these leaping quanta form. These patterns are bound to be local, by Eberhard's proof, and have never been observed to be otherwise. On the other hand the quanta themselves—the unpredictable alphabet which spells out the words and paragraphs of the world's phenomena—must be non-locally connected, according to the theorem of John Bell.

Quantum Reality #3: The world is an undivided wholeness. The notion that the world is an inseparable whole arises from the presence in quantum theory of "phase entanglement." In the quantum formalism, two quons that have once interacted do not separate into two waveforms when they move apart but are forever afterward represented as a single wave. Whether this *wholeness of representation* is matched by a *wholeness of being* is a question that was posed by certain thoughtful physicists, especially Erwin Schrödinger and David Bohm.

Although superluminal phase entanglement is necessary to make the answers come out right, it never leads to any superluminal results. Since these instant connections which bind separated quons into a seamless whole (in the formalism, at least) never surface in the world of phenomena, most physicists regarded them as purely formal features of the quantum language—necessary for calculation but having no counterpart in reality.

Bell's theorem shows that the holistic grammar of the quantum formalism reflects the inseparable nature of reality itself. Beneath phenomena, the world is a seamless whole.

Although it points beyond phenomena, Bell's theorem is proved by arguments drawn solely from the facts. Because of its strictly phenomenal base, Bell's theorem by itself gives no hint as to the mechanism by which reality might achieve its necessarily non-local connections. These ubiquitous phase connections in the quantum formalism offer a non-classical image for how a non-local world might work: quons are instantly connected not because something stretches between them but because each has left part of itself in the other, a part to which it retains immediate access.

Quantum Reality #4: The many-worlds interpretation. Bell's theorem has nothing to say directly about the many-worlds interpretation because in Everett's luxurious universes you cannot prove Bell's theorem.

The subtle but necessary CFD (contrafactual definiteness) assumption takes for granted that for each photon pair in the EPR experiment a particular hypothetical calcite setting will lead to a definite result. In the many-worlds model of reality, each measurement setting leads to *all possible results.* The Everett multiverse violates the CFD assumption because although such a world has plenty of contrafactuality, it is short on definiteness.

Although Bell's theorem does not apply to an Everett-style universe, there's plenty of non-locality present without it. Any model of reality in which a tiny event in the Andromeda galaxy can instantly split my reality into thousands of Xerox copies cannot by any stretch of the imagination be called "local."

Quantum Reality #5: Quantum logic (The world is put together like a non-Boolean lattice). To cover the facts takes more than logic. Quantum logic, like its Boolean cousin, codifies just the bare bones of talk about quantum attributes. It provides not a complete quantitative picture of quantum phenomena, but a mere logical skeleton which needs to be fleshed out by more specific quantitative relations.

Bell's proof cannot be derived from qualitative arguments but depends for its validation on specific numerical relationships. See Fig. 12.5 for an illustration of just how small an experimental difference separates a non-local reality from its local competition. Quantum logical relations merely outline the phenomena and do not provide specific quantitative information. Bell's theorem can't be proved from logic lattices alone.

It has been suggested that even though one cannot *prove* Bell's theorem via quantum logic, perhaps one can *disprove* it. To prove this theorem, you must have recourse to ordinary Boolean logic; if the world's logic is otherwise, perhaps this proof doesn't go through. However, even quantum logicians use Boolean logic when it comes to *talking about* quantum logic. In other words, the *metalogic* of quantum logic is Boolean.

In addition, quantum logic is the consequence of a particular assumption about a quon's attributes. If you assume that photons have attributes of their own, then these attributes must follow a non-Boolean arithmetic. On the other hand, if you assume that a photon's attributes are relational, ordinary logic suffices. Since the version of Bell's theorem presented here makes no assumptions at all about photon attributes, the concept of quantum logic is irrelevant.

Bell's theorem likewise has not provided new insight into the meaning of non-Boolean relations. Of all quantum realities, Bell's logical necessity of non-local connections seems to illuminate quantum logic the least.

Quantum Reality #6: Neorealism (The world is made of ordinary objects). Von Neumann's proof outlaws objects with innate attributes. Bell's theorem also forbids such ordinary objects and in addition rejects all entities whose *relational attributes* depend only on the settings of *local M devices.* If these proofs are valid, the only kinds of entities which can make up the quantum world are those whose attributes depend *non-locally* on the settings of distant measuring devices.

Most neorealist models consider the world to be built in the old-fashioned classical manner, out of *particles* and *fields.* Models of quantum reality constructed according to this plan possess the peculiar feature that the fields which connect each particle to its environment seem to be capable of instantly switching a particle's attributes in response to a configuration change anywhere in the universe.

Such particle/field models of the world are not unmediated, because changes in a particle's attributes are carried by a field rather than mysteriously jumping from one point to another. However, this field—called the pilot wave—must be able to transmit information faster than light without attenuation. Thus although these neorealist worlds are mediated (by pilot waves), they are still *unmitigated* and *immediate*.

For a physicist the most unpleasant feature of these neorealist models is the presence of *real fields* that move faster than light. Bell's theorem shows that the appearance in neorealist schemes of fields capable of superluminal data transmission is no accident: any model of reality which fits the quantum facts must possess some means of exchanging information faster than light.

Determined foes of superluminal connections cannot agree which option is more repulsive: the real superluminal fields of the neorealists, which explicitly carry signals from place to place at FTL velocities, or the voodoo-like unmediated influences suggested by the quantum formalism, which simply jump directly from Earth to Betelgeuse.

Quantum Reality #7: Consciousness creates reality. Bell's theorem says that if consciousness does indeed create reality, it cannot be a purely local matter. The decision of a mind *here* must be able to change the value of an attribute *there*, where *here* and *there* may be separated by immense distances.

Rash speculations that this strong quantum connection permits telepathy or long-distance mind-over-matter effects need to be balanced by the realization that Bell's theorem concerns only reality—that is, raw quantum jumps—not phenomena, the regular patterns of quantum jumps. Thus even if consciousness could create reality, the power of mind to intervene in distant happenings may be limited to the production of single, statistically unusual events—the so-called outriders or glitches that show up from time to time in even the most well-controlled experiment.

That faraway minds can alter the fabric of reality but not the pattern woven thereon need not limit minds to the production of trivial deeds. Some physicists believe that the whole physical universe originated out of nothing as a single quantum jump—just the sort of wild, unpredictably unique quantum event that a mind could initiate without upsetting the statistical applecart.

Quantum Reality #8: The duplex universe. Werner Heisenberg was one of the few quantum physicists who tried to imagine what unmeasured quantum reality might look like. According to Heisenberg, the world sans M devices is not fully real but consists of a *superposition*—a particularly intimate quantum style of coexistence—of half-real "tendencies for being" which he called *potentia.* The advantages of such an attenuated style of being is that many contradictory tendencies can coexist, an option not open to solid facts; the price of non-contradiction is that none of these tendencies is completely "real." Upon measurement, but not before, one of these tendencies is selected, apparently at random, from the flock, and promoted to full reality status. The essence of measurement, in Heisenberg's duplex world, is the sudden transformation of potentia into actuality.

Bell's theorem requires that this measurement-induced transition from soft possibility to hard actuality cannot be local but must depend on other measurements going on at locations arbitrarily distant.

Heisenberg's model is an unusually explicit version of observer-created reality. Unlike other proponents of observer-created reality, he tries to imagine what reality is like before observation. Heisenberg declares the raw material of the universe to be potentia, tendency, possibility   a world, in a word, founded on a wave of opportunity. Bell's theorem applied to Heisenberg's picture requires that these oscillating opportunities be linked together faster than light.

This brief review of the eight quantum realities in the light of Bell's proof shows that Bell's theorem does not resolve the quantum reality question in favor of one reality or another. As long as they leave room for non-locality, all eight of these realities are viable candidates for a model of "the way the world really is."

What Bell's theorem does do for the quantum reality question is to clearly specify one of deep reality's necessary features: whatever reality may be, *it must be non-local.* Since Clauser's experimental verification of Bell's theorem, we know that any correct model of reality has to incorporate explicit non-local connections. No local reality can explain the type of world we live in.

Furthermore, since Bell's result is based on experimental facts, it is independent of whether quantum theory is correct or not. Should quantum theory someday fail in its predictions or simply be replaced by an entirely different way of predicting the same quantum facts, Bell's theorem would still be valid. Although it arose in the context of disputes about the completeness of quantum theory, Bell's theorem is derived from the facts themselves, not from any particular theoretical representation of these facts.

Bell's theorem has illuminated one corner of deep reality, but the reality crisis in physics is far from over; as yet no physicist can tell you what sort of world we happen to live in. I speculate next on a few directions the quest for quantum reality might take.

CAN WE DEVISE AN EXPLICITLY HOLISTIC MODEL OF MEASUREMENT?

One of the most artificial features of quantum theory is its division of the world into two parts: system and M device. The quantum formalism implies that the world is a seamless whole, yet the first step in any quantum computation is to fracture that unity.

In the all-quantum interpretation of measurement a von Neumann chain stretches between system and observer, and the measurement problem consists of where to "break" that chain. In light of quantum wholeness, representing a measurement as a linear chain with definite beginning and end seems like a bad way to start. Perhaps the von Neumann chain, rather than being broken, should be welded together into a loop to form what we might call a von Neumann ring.

In such a scheme (admittedly vague), the measuring device would provide a context for the quantum system's attributes while at the same time its constituent quantum systems would provide a context for the attributes of the M device. Each quantum entity would measure the other, closing the circle and achieving a self-contained and consistent model of the measurement process.

This vision of systems measuring one another would allow for the existence of autonomous realities without the need for human observers. In a world consisting of communities of mutually scrutinizing measurement loops, few quons would not belong to at least one such community. Even those entities that have ceased to interact conventionally with their fellows would be implicated in one or more von Neumann rings via non-local phase entanglement.

This hypothetical reality made up of von Neumann rings takes Bohr's view to the limit: it is thoroughly *relational*, containing no privileged entities of any kind; a world entirely thingless, full of quons like rainbows which are themselves made of rainbows.

Another possible direction for reality research would be the actual exploration of the multiple worlds suggested by Everett's interpretation of quantum theory.

According to Kant, humans cannot experience reality itself because our senses and brains were developed for more mundane purposes. For instance, one of the biggest lessons that Einstein's special theory of relativity teaches us is that, in reality, this world is *four-dimensional*. Although humans experience a three-dimensional world in which time seems to flow, reality itself exists in space-time—in which time is a dimension on a par with space. Hermann Minkowski, one of Einstein's colleagues, introduced the four-dimensional structure of the world this way to his students:

"The views of space and time which I wish to lay before you have sprung from the soil of experimental physics, and therein lies their strength. They are radical. Henceforth space by itself, and time by itself, are doomed to fade away into mere shadows, and only a kind of union of the two will preserve an independent reality."

If time is just another dimension, then the entire history of the universe from beginning to end is spread along this time line. The past still exists and so does the future. Our human perception of an eternal present which seems to travel along in the future direction is an illusion: that's not the way the world is at all. Physicist Hermann Weyl expressed the four-dimensional view of things this way: "The objective world simply *is;* it does not *happen.*"

With what authority do physicists deny their immediate experiences to claim that the world's really four-dimensional? Just this: if they don't write the equations of physics in four dimensions (and instead write them in three space dimensions plus a changing temporal dimension), they get answers that agree with the experience of stationary observers but not with observers moving relative to them. To accurately describe electrical and mechanical phenomena in a manner valid for all observers, physicists have to visualize the world in space-time, not in space plus time.

Likewise, in quantum theory physicists must describe the unmeasured world as a simultaneous superposition of all its possibilities at once. If they leave out a single possibility, they get the wrong answer. However, we do not *experience* the world as a superposition of possibilities, but only as a one-at-a-time sequence of definite actualities. Once again human perception of things seems out of step with reality.

No matter how sophisticated our concepts, we cannot help but perceive the world, according to Kant, except through particularly human filters. Even if we knew better, we couldn't tear off our colored spectacles and look at the world as it really is. In this Kantian spirit Heisenberg describes how phenomena inevitably appear classical to humans (Cinderella effect) despite signals from physics that reality itself is anything but classical:

"Any experiment in physics, whether it refers to the phenomena of daily life or to atomic events, is to be described in the terms of classical physics. The concepts of classical physics form the language by which we describe the arrangement of our experiments and state the results. We cannot and should not replace these concepts by any others . . . The use of classical concepts is finally a consequence of the general human way of thinking . . . There is no use in discussing what could be done if we were other beings than we are."

The source of all quantum paradoxes appears to lie in the fact that human perceptions create a world of unique actualities—our experience is inevitably "classical"—while quantum reality is simply not that way at all. Quantum reality consists of simultaneous possibilities, a "polyhistoric" kind of being absolutely incompatible with our merely one-track minds.

If these alternative universes are really real and we are barred from experiencing them only by a biological accident, perhaps we can mechanically extend our senses—as we have in so many other cases—with a sort of quantum "microscope" which would allow us to actually experience some of Everett's parallel universes firsthand. Since physics assures us that our lives are embedded in a thoroughly quantum world, is it so obvious that our experiences must remain forever classical?

IS CONSCIOUSNESS A TYPE OF QUANTUM KNOWLEDGE?

Although it seems to be true that every physics experiment is classical in form if not in content, is it so obvious that the full range of human experience is also absolutely classical?

Science's biggest mystery is the nature of consciousness. It is not that we possess bad or imperfect theories of human awareness; we simply have no such theories at all. About all we know about consciousness is that it has something to do with the head, rather than the foot. That's not much but it appears to be more than the ancient Egyptians knew: the Egyptians threw away the brain before beginning their elaborate embalming procedures, judging it to be a mere accessory.

Is it possible that consciousness is some sort of quantum effect? Is human awareness a privileged access to the "inside" of the quantum world, an open door to some brain quon's realm of possibility? Can we know firsthand what it is like to dwell in the quantum world just by sitting still and looking inside our heads?

Human mental experience seems to be of two kinds—an experience of facts, memories, emotions, body states—a thoroughly classical kind of knowing which we might call "computer consciousness," which takes place against a peculiar background of "raw awareness"—that uncanny yet familiar feeling we relinquish when we go to sleep and awaken into every morning. Some have called this second kind of experience "consciousness without an object." I call it "ordinary awareness" and believe that it is one human quality that distinguishes us from computers—at least computers as they are presently constituted.

If ordinary awareness is a direct connection to quantum reality, then just as our *external* knowledge of quantum entities may be characterized by the term "quantum ignorance," so we might call this immediate *internal* experience of the world's real nature "quantum knowledge." One of the greatest scientific achievements imaginable would be the discovery of an explicit relationship between the waveform alphabets of quantum theory and certain human states of consciousness.

Bell's theorem shows that although the world's phenomena seem strictly local, the reality beneath this phenomenal surface must be superluminal. The world's deep reality is maintained by an invisible quantum connection whose ubiquitous influence is unmediated, unmitigated, and immediate. Unconfirmed rumors of telepathy and other alleged powers of mind aside, our basic computer consciousness appears to be as local as any other classical phenomenon. However, if ordinary awareness is a private manifestation of deep quantum reality, Bell's theorem *requires* our quantum knowledge to be non-local, instantly linked to everything it has previously touched. Since this type of awareness consists of consciousness *without content*, it is difficult to see what use we could make of such non-local connections. On the other hand, perhaps these connections are not there for us to "use."

Religions assure us that we are all brothers and sisters, children of the same deity; biologists say that we are entwined with all life-forms on this planet: our fortunes rise or fall with theirs. Now, physicists have discovered that the very atoms of our bodies are woven out of a common superluminal fabric. Not merely in physics are humans out of touch with reality; we ignore these connections at our peril. Albert Einstein, a seeker after reality all his life, had this to say concerning the illusion of separateness:

"A human being is part of the whole, called by us 'Universe'; a part limited in time and space. He experiences himself, his thoughts and feelings as something separated from the rest—a kind of optical delusion of his consciousness. This delusion is a kind of prison for us, restricting us to our personal desires and to affection for a few persons nearest us. Our task must be to free ourselves from this prison by widening our circle of compassion to embrace all living creatures and the whole of nature in its beauty. Nobody is able to achieve this completely but the striving for such achievement is, in itself, a part of the liberation and a foundation for inner security."

APPENDIX I

Additional Readings on Quantum Reality

The following list of books and articles on the quantum reality question is not meant to be complete but represents material I have found useful in making sense of this field.

GENERAL REFERENCES

1. *Foundations of Quantum Mechanics: Proceedings of the International School of Physics "Enrico Fermi" Course 49*, Bernard d'Espagnat ed. New York: Academic Press (1971).

2. *The Philosophy of Quantum Mechanics*, Max Jammer. New York: Wiley (1974).

3. *Conceptual Foundations of Quantum Mechanics (Second Edition)*, Bernard d'Espagnat. Reading, Mass.: W. A. Benjamin (1976).

4. *Quantum Theory and Measurement*, John Archibald Wheeler & Wojciech Hubert Zurek eds. Princeton, N.J.: Princeton University Press (1983).

These four books are essential reading for serious reality researchers. Reference #1 is a record of a summer school on quantum foundations; Reference #2 is a detailed historical review of the quantum reality question. The book by d'Espagnat is the nearest thing to a textbook in this field. Reference #4 is a collection of important articles on quantum reality research; articles cited here that are also reprinted in the Wheeler-Zurek collection are marked with an asterisk.

SPECIFIC QUANTUM REALITIES

5. *Atomic Physics and Human Knowledge*, Niels Bohr. New York: Wiley (1963).
6. "The Copenhagen Interpretation," Henry Stapp. *American Journal of Physics* 40 1098 (1972).
7. "Law Without Law" (*), John Archibald Wheeler (see Wheeler & Zurek, Ref. #4, p. 182).
8. *Wholeness and the Implicate Order*, David Bohm. London: Routledge and Kegan Paul (1980).
9. *The Many-Worlds Interpretation of Quantum Mechanics*, Bryce DeWitt & R. Neill Graham. Princeton, N.J.: Princeton University Press (1973).
10. "The Logic of Quantum Physics," David Finkelstein. *Transactions of the NY Academy of Sciences 25* #6 621 (1965).
11. "Quantum Logic," Carl G. Adler & James F. Wirth. *American Journal of Physics 51* 412 (1983).
12. "Remarks on the Mind-body Question" (*) in *The Scientist Speculates*, I. J. Good, ed.; New York: Basic Books (1962).
13. "Mind, Matter and Quantum Mechanics," Henry P. Stapp. *Foundations of Physics 12* 363 (1982).
14. *A Survey of Hidden-Variables Theories*, Frederik J. Belinfante. Oxford: Pergamon Press (1973).
15. "Measurement Understood through the Quantum Potential Approach," David Bohm and Basil Hiley. *Foundations of Physics 14* 255 (1984).
16. *Physics and Philosophy*, Werner Heisenberg. New York: Harper & Brothers (1958).

These are the best references I can recommend on the eight quantum realities described in the text.

TEXTBOOKS OF PARTICULAR INTEREST

17. *Mathematical Foundations of Quantum Mechanics* (English translation by R. T. Beyer), John von Neumann. Princeton, N.J.: Princeton (1955).
18. *Quantum Theory*, David Bohm. New York: Prentice-Hall (1951).
19. *The Feynman Lectures on Physics: Volume III*, Richard P. Feynman, Robert B. Leighton, Matthew Sands. Reading, Mass.: Addison-Wesley (1965).
20. *Quantum Mechanics and Path Integrals*, R. P. Feynman and A. R. Hibbs. New York: McGraw-Hill (1965).

Reference #17 is von Neumann's "quantum bible"—the mathematical framework which still holds up quantum theory. Bohm's textbook is a clear presentation of quantum theory from the Copenhagen standpoint written before his defection to the neorealist camp. Reference #19 is an unusually lucid introduction to this theory. In Reference # 20, Feynman describes his sum-over-histories approach in considerable detail.

THE EPR PARADOX AND BELL'S THEOREM

21. "Can Quantum-mechanical Description of Physical Reality Be Considered Complete?" (*), Albert Einstein, Boris Podolsky, Nathan Rosen. *Physical Review 47* 777 (1935).
22. "Can Quantum-mechanical Description of Physical Reality Be Considered Complete?" (*), Niels Bohr. *Physical Review 48* 696 (1935).
23. "On the Problem of Hidden Variables in Quantum Mechanics" (*), John S. Bell. *Reviews of Modern Physics 38* 447 (1966).
24. "On the Einstein-Podolsky-Rosen Paradox" (*), John S. Bell. *Physics 1* 195 (1964).
25. "Bell's Theorem: Experimental Tests and Implications," John F. Clauser and Abner Shimony. *Reports on Progress in Physics 41* 1881 (1978).
26. "Experimental Test of Bell's Inequalities Using Time-varying Analyzers," Alain Aspect, Jean Dalibard, Gerard Roger. *Physical Review Letters 49* 1804 (1982).

Reference # 21 is the original EPR paper and # 22 is Bohr's reply. Reference # 23 is Bell's definitive analysis of von Neumann's proof and his premonitions of Bell's theorem. Reference # 24 is Bell's original formulation of his celebrated theorem. The Clauser-Shimony article is a recent review of the experimental status of Bell's theorem. The last paper is a description of Aspect's verification of Bell's theorem using ultrafast polarization switches.

SPECIAL TOPICS

27. *The Ethereal Aether,* Loyd S. Swenson, Jr. Austin: University of Texas (1972).
28. *Molecular Reality,* Mary Jo Nye. New York: American Elsevier (1972).
29. *The Intensity Interferometer,* Robert Hanbury Brown. New York: Halsted Press (1974).
30. "Quantum Non-Demolition Measurements" (*), V. B. Braginsky, Y. I. Vorontsov, K. S. Thorne. *Science 209* 547 (1980).
31. "Squeezed States of Light," D. F. Walls. *Nature 306* 141 (1983).

The first two references chronicle the rise and fall of the luminiferous ether and the ascent of the atomic hypothesis into indubitable reality status. The remaining articles provide more details on the widths of photon proxy waves and the unusual quantum attributes which are the subject of so-called QND measurements.

SUPERLUMINAL SIGNALING

32. "FLASH—A Superluminal Communicator Based upon a New Kind of Quantum Measurement," Nick Herbert. *Foundations of Physics 12* 1171 (1982).
33. "Bell's Theorem and the Different Concepts of Locality," Philippe Eberhard. *Nuovo Cimento 46B* 392 (1978).
34. "A Single Quantum Cannot be Cloned," W. K. Wooters and W. H. Zurek. *Nature 299* 802 (1982).
35. "Is a Photon Amplifier Always Polarization Dependent?" L. Mandel. *Nature 304* 188 (1983).

Reference #32 is a typical proposal purporting to use quantum connectedness to signal faster-than-light. Reference #33 contains Eberhard's proof that such proposals must fail if quantum theory is correct and complete. References #34 and #35 supply the detailed refutation of this particular signaling scheme.

POPULARIZATIONS OF QUANTUM REALITY RESEARCH

36. *The Tao of Physics*, Fritjof Capra. Berkeley, Calif.: Shambhala (1975).
37. *Fabric of the Universe*, Denis Postle. New York: Crown (1976).
38. *The Dancing Wu Li Masters*, Gary Zukav. New York: Morrow (1979).
39. *Other Worlds*, Paul Davies. New York: Simon & Schuster (1980).
40. *Taking the Quantum Leap*, Fred Alan Wolf. New York: Harper & Row (1981).
41. *The Cosmic Code*, Heinz R. Pagels. Simon & Schuster (1982).
42. *In Search of Reality*, Bernard d'Espagnat. Berlin, West Germany: Springer-Verlag (1983).
43. *The Quantum World*, J. C. Polkinghorne. Harlow, Essex, England: Longman (1984).
44. *In Search of Schrödinger's Cat*, John Gribbin. New York: Bantam (1984).

These presentations of the quantum reality question illustrate the divergent and contradictory views of physicists (and their interpreters) concerning the nature of deep reality. Of particular interest is Pagels' "reality marketplace" in Reference #41.

ARTICLES ON QUANTUM REALITY

45. "Quantum Mechanics and Reality," Bryce DeWitt. *Physics Today 23* p. 30 September (1970).
46. "Quantum Theory and Reality," Bernard d'Espagnat. *Scientific American 241* p. 158 November (1979).

47. "Ghostly Interactions in Physics," Basil Hiley. *New Scientist* p. 746 March 6 (1980).
48. "Physicists Redefine Reality." London *Economist* p. 95 September 26 (1981).
49. "Quantum Mysteries for Everyone," N. David Mermin. *Journal of Philosophy 78* 397 (1981).
50. "Quantum Weirdness," Martin Gardner. *Discover 3* #10 69 (1982).
51. "Quantum Mechanics Passes Another Test," Arthur L. Robinson. *Science 217* 435 (1982).
52. "Loophole Closed in Quantum Mechanics Test," Arthur L. Robinson. *Science 219* 40 (1983).
53. "Facing Quantum Mechanical Reality," Fritz Rohrlich. *Science 221* 1251 (1983).

These articles are mainly for the benefit of scientists working in fields other than quantum reality research or for the general public. The report on scientists' redefinition of reality (Reference #48) in the London *Economist* appears to have made no impact whatsoever on the world's financial reality.

Direct Experience

54. *Reality and Empathy*, Alex Comfort. Albany: State University of New York (1984).
55. *The Sex Sphere*, Rudy Rucker. New York: Ace Science Fiction (1983).
56. *Mr. Tompkins in Paperback*, George Gamow. Cambridge, England: Cambridge University Press (1965).
57. *Cosmicomics*, Italo Calvino. New York: Harcourt, Brace & World (1968).
58. *All the Myriad Ways*, Larry Niven. New York: Ballantine (1971).
59. *Bell's Book*, Helen Luster. Fur Line Press (1976): available from Manroot Books, Box 982, South San Francisco, CA 94080.
60. *Schrödinger's Cat*, Robert Anton Wilson. New York: Pocket Books (1981).
61. *Superluminal*, Vonda McIntyre. Boston: Houghton Mifflin (1983).

Delving into the mysteries of everyday experience, Alex *(Joy of Sex)* Comfort interviews the demon "Gezumpstein" who can perceive the world "as it really is." Rudy Rucker's like-minded demon calls herself "Babsi."

In Mr. Tompkins' world, new values for the physical constants make quantum and relativity effects part of everyday life. Professor Gamow I'm sure meant these imaginary worlds to be taken with a grain of salt, not as serious pictures of quantum reality. *Cosmicomics* explores what a consciousness-centered cosmos might feel like from the inside. Niven's story describes some unexpected psychological consequences of the many-worlds reality. Using poetic reverie as probe, Luster's *Bell's Book* tests our linguistic safety net for irregular connections. Various quantum realities form the backdrop for Wilson's wacky freak show. In McIntyre's *Superluminal*, the achievement of faster-than-light travel opens up new space-time possibilities for humankind.

Quantum Number

BELL'S THEOREM BLUES

Words: Nick Herbert   Music: Traditional Blues
(Number Twelve Train)

Doc-tor Bell say we con - nec-ted.

He call me on the phone.

Doc-tor Bell say u - ni-ted.

He call me on the phone.

But if we re-al-ly to gether ba - by

How come I feel so all a - lone?

© 1984 Nick Herbert

INDEX